Math Series
ADVANCED

ALGEBRA

by Stephen B. Jahnke

Book cover design by Kathy Kifer

Dedicated
to Joy Leal
for her help and insight.

Published by
Garlic Press
100 Hillview Lane #2
Eugene, OR 97401

ISBN 0-931993-37-7
Order Number GP-037

CONTENTS

Pre-Algebra Review

In this book we will assume that the student is already familiar with certain pre-algebra skills. A few topics are listed below for reference. For a more detailed explanation of these skills consult the Straight Forward Math Series Pre-Algebra Book 1 and Book 2 or any similar text on pre-algebra.

Commutative Property

The order in which numbers are added or multiplied does not change the answer.

•Examples:

$3 + 5 = 5 + 3$	Both sides equal 8.
$2 \cdot 7 = 7 \cdot 2$	Both sides equal 14.

Associative Property

The way in which three numbers are grouped with parentheses in addition or multiplication does not change the answer.

•Examples:

$(1 + 2) + 3 = 1 + (2 + 3)$	Both sides equal 6.
$(2 \cdot 3) \cdot 5 = 2 \cdot (3 \cdot 5)$	Both sides equal 30.

Identity Property of Addition

A number plus zero is always itself.

•Example:

$$11 + 0 = 11$$

Identity Property of Multiplication

A number multiplied by 1 is always itself.

•Example:

$$17 \cdot 1 = 17$$

•Further examples:

$5 + x = x + 5$	Commutative Property of Addition
$(x+2) + 7 = x + (2+7)$	Associative Property of Addition
$y + 0 = y$	Identity Property of Addition
$x \cdot 3 = 3 \cdot x$	Commutative Property of Multiplication
$4 \cdot (2 \cdot y) = (4 \cdot 2) \cdot y$	Associative Property of Multiplication
$1 \cdot z = z$	Identity Property of Multiplication

Opposites

The opposite of a number is found by changing its sign.

•Examples:

4 is the opposite of -4

-7 is the opposite of 7

Addition with Negative Numbers

Case 1: If both numbers are negative, add the numerals and write a negative in front of the result.

•Examples:

(-2) + (-9) = -11
(-3) + (-13) = -16

Case 2: If only one number is negative, subtract the smaller numeral from the larger numeral. Write a negative in front of the result <u>only</u> if the larger numeral was negative.

•Examples:

(-4) + (5) = 1
-10 + 7 = -3
9 + (-1) = 8
9 + (-11) = -2

Subtraction with Negative Numbers

Replace the subtraction symbol with an addition symbol and also change the second number to its opposite. Then follow the cases in addition above.

•Examples:

(-2) - (9) = (-2) + (-9) = -11	As in Case 1.
(-4) - (-5) = (-4) + (5) = 1	As in Case 2.
-10 - (-7) = -10 + 7 = -3	As in Case 2.
9 - 11 = (9) - (11) = 9 + (-11) = -2	As in Case 2.
-3 - 13 = -3 - (13) = -3 + (-13) = -16	As in Case 1.

Multiplication and Division with Negatives

If the signs are the same, the answer is positive. If the signs are different, the answer is negative.

•Examples:

Same Signs	Different Signs
$(-3)(-10) = 30$	$(7)(-3) = -21$
$(2)(5) = 10$	$(-4)(5) = -20$
$-15 \div -3 = 5$	$\dfrac{8}{-2} = -4$
$\dfrac{-14}{-7} = 2$	$\dfrac{-6}{3} = -2$

Order of Operations

Calculate in this order:

1st - Parentheses
2nd- Exponents
3rd- Multiply and divide (from left to right).
4th- Add and subtract (from left to right).

•Example:

Calculate	$18 \div (2+1)^2 \cdot 4 - 5$	
	$18 \div 3^2 \cdot 4 - 5$	Parentheses
	$18 \div 9 \cdot 4 - 5$	Exponent
	$2 \cdot 4 - 5$	Divide
	$8 - 5$	Multiply
	3	

Review of Pre-Algebra Skills. Exercise 1.
Name the following properties:

1. $2 + 3 = 3 + 2$

2. $11 \cdot 5 = 5 \cdot 11$

3. $7+(6+3) = (7\ 6)+3$

4. $-2 \cdot (4 \cdot 3) = (-2 \cdot 4) \cdot 3$

5. $17 + 0 = 17$

6. $-14 \cdot 1 = -14$

7. $4+x = x+4$

8. $-3+(5+x) = (-3+5)+x$

9. $y \cdot 1 = y$

10. $5 \cdot (x \cdot y) = (5 \cdot x) \cdot y$

11. $0 + z = z$

12. $-2 \cdot z = z \cdot (-2)$

13. $a+b = b+a$

14. $a+(b+c) = (a+b)+c$

15. $a+0 = a$

16. $a \cdot b = b \cdot a$

17. $a \cdot (b \cdot c) = (a \cdot b) \cdot c$

18. $a \cdot -1 = a$

19. Find the opposite of:

        a. 4   b. -2   c. 11   d. -30   e. 0

Calculate:

20. $-4 + (-8)$

21. $2 + (-7)$

22. $-5 + 6$

23. $1 - 5$

24. $-3 - 4$

25. $3 - (-2)$

26. $-2 - (-6)$

27. $7 - 4$

28. $-7 - 4$

29. $7 - (-4)$

30. $-7 - (-4)$

31. $(-3) (-6)$

32. $(5) (-2)$

33. $22 \div (-2)$

34. $\dfrac{-10}{5}$

Calculate:

35. $2 \cdot 3 + 2 \cdot 5$

36. $2 \cdot (3 + 5)$

37. $(-3) (-2) + 5^2$

38. $4 - 2 + 5 \cdot 3^2$

39. $5 + 2 \cdot (3+1)^2$

40. $2 - 8 + (3-1)^2$

5

Distributive Law

The distributive property (also called the Distributive Law) states: For any three numbers a, b, and c, it is true that $a(b+c) = ab + ac$.

•Example:

$$2(3 + 4) = 2 \cdot 3 + 2 \cdot 4$$ We can work both sides of the equation to show this is true:

Left Side	Right Side
$2(3 + 4)$	$2 \cdot 3 + 2 \cdot 4$
$2(7)$	$6 + 8$
14	14

Both sides are 14. Hence $2(3 + 4) = 2 \cdot 3 + 2 \cdot 4$.

In other words, the number 2 outside the parentheses (on the left side of the equal sign) has been *distributed* to both the 3 and the 4 (on the right side of the equal sign).

Distributive Law. Exercise 1. Work out the left side separately from the right side to convince yourself that the Distributive Law is true.

1. $7(2 + 3) = 7 \cdot 2 + 7 \cdot 3$
2. $8(1 + 2) = 8 \cdot 1 + 8 \cdot 2$
3. $100(2 + 7) = 100 \cdot 2 + 100 \cdot 7$
4. $1 \cdot (4 + 6) = 1 \cdot 4 + 1 \cdot 6$

5. $3(-2 + 5) = 3 \cdot (-2) + 3 \cdot 5$
6. $(2 + 3) \cdot 4 = 2 \cdot 4 + 3 \cdot 4$
7. $(7 + 3) \cdot 2 = 7 \cdot 2 + 3 \cdot 2$
8. $(-2 + 4) \cdot 3 = -2 \cdot 3 + 4 \cdot 3$

Distributive Law. Exercise 2. Fill in the right side of each equation by distributing the number outside the parentheses to each number inside the parentheses.

1. $4(3 + 7) = 4 \cdot 3 + 4 \cdot 7$
2. $6(1 + 2) =$
3. $5(5 + 4) =$
4. $9(x + 2) =$
5. $7(y + 1) =$
6. $2(3 + m) =$
7. $5(x + y) =$
8. $3(-1 + 6) =$

9. $12(-2 + 3) =$
10. $4(-x + 3) =$
11. $(7 + 2)3 = 7 \cdot 3 + 2 \cdot 3$
12. $(1 + 6)2 =$
13. $(-2 + 7)4 =$
14. $(-12 + 5)7 =$
15. $(x + 3)4 =$
16. $a(b + c) =$

Operations, Expressions, Terms and Like Terms

Addition (+), subtraction (-), multiplication (·) and division (÷) are called **operations**.

An **expression** is any collection of numbers, variables, parentheses or operations.

For example: **-5x - 10y + 4y + x** is an expression.

The addition and subtraction symbols separate this expression into four parts called **terms**. The terms in this expression are *-5x* , *-10y* , *+4y* , and *+x*. In other words, the negative sign in front of the *5x* together with the 5x makes the term *-5x*. The negative sign in front of the *10y* together with the *10y* makes the term *-10y* . Likewise, the positive sign in front of the *4y* together with the *4y* makes the term *+4y* . The positive or negative sign is part of the term; however, many times the plus sign is not written down. For instance: *+4y* is often written as just *4y*.

Like terms are terms of the same letter. In **-5x - 10y + 4y + x**, the *-10y* and *+4y* are like terms, because they are both *y* terms. The *x* like terms are *-5x* and *+x* .

Coefficients are number parts (including the signs) of each term. The coefficient of *-5x* is *-5* . The coefficient of *-10y* is *-10* . The coeffecent of *4y* is *4*. If a term has no number component, as in the term *x* , then the coefficient is assumed to be *1*. This is because $x = 1 \cdot x$ (Identity Property).

•Example 1:

x - y	
Terms:	x and -y
Coefficients:	1 and -1, because $x = 1 \cdot x$ and $-y = -1 \cdot y$.
Like terms:	None

•Example 2:

8y - 5 + 7	
Terms:	8y, -5, and +7
Coefficients:	8, -5, and +7
Like Terms:	-5 and +7, because neither term contains a variable.

•Example 3:

```
-z + 8 + 6z
        Terms:              -z, 8, 6z
        Coefficients:       -1, 8, 6, since -z = -1 · z  = -1z
        Like terms:         -z and 6z
```

Distributive Law. Exercise 3. For each of the following, list the terms, list the coefficients, and list the like terms.

1. -5x - 3x- 2y - 7y	5. 10x - y	9. 4x + 5 + 8
2. 7y - 2x + 14x	6. x - x	10. -3 + 2 + 0 + z
3. 4z + x - 6x	7. 9 - 6	
4. 2x + 3y + 4z + 5	8. -89x	

Collecting Like Terms

To collect like terms, add the coefficients of the like terms together followed by the common letter.

Examples: Reason:

2x + 3x = 5x	2 + 3 = 5
-7y + 10y = 3y	-7 + 10 = 3
5x - 5x = 0x = 0	5 + (-5) = 0 and 0 · x = 0, because 0 · (anything) = 0
y + 2y = 3y	y + 2y = 1y + 2y = 3y
-x + 10x = 9x	-x + 10x = - 1x + 10x = 9x
y - y = 0	y - y = 1y - 1y = 0y = 0
5x + 8y - y - 4x - x = 7y	8y - y = 7y and 5x - 4x - x = 0x = 0

Distributive Law. Exercise 4. Collect like terms.

1. 2x + 7x

2. 14y - 4y

3. 10d + d

4. 8y - 8y

5. 5x -7x +3y +4y

6. $\frac{1}{3}$x + $\frac{1}{3}$x

7. -x + 3x + 5 (Remember: -x = -1x)

8. 7y - y + 4y (Remember: -y = -1y)

9. y - y + 2x -2x

10. 4x + 3y + 2p + 6x

11. -3x -10x - 4y - 2y

12. 1.25x + 2.75x

13. 5 - 4x +2

14. 4a + 5 - 5a +4

15. 13x + 5a + 7y + 6a + x

16. 5x + 6x + 2x + 10y - 10y

Distributive Law in General

Multiply the number outside the parentheses to each term in the parentheses. Pay close attention to the sign on each term.

•Example:

Simplify -4(3x - 2).

The terms in the parentheses are *3x* and *-2* . Remember the negative in front of the 2 makes the term *-2* . Multiplying the *-4* to *3x* and *-2* we get:

-4 · 3x = -12x

-4 (3x - 2) =   - 12x + 8

-4 (-2) = +8

The paths above are usually done mentally and the entire process looks like this:

-4(3x - 2) = -12x + 8

Reasons

$-3(x + 5) = -3x - 15$ $-3 \cdot x = -3x$ and $-3(+ 5) = -15$

$2(x - 7) = 2x - 14$ $2 \cdot x = 2x$ and $2(-7) = -14$

$5(2y + 1) = 10y + 5$ $5 \cdot 2y = 10y$ and $5(+ 1) = +5$

$6(-3x - 4) = -18x - 24$ $6(-3x) = -18x$ and $6(-4) = -24$

$-2(-3p - 4) = 6p + 8$ $-2(-3p) = 6p$ and $-2(-4) = +8$

$-4(-x + 3m - 1) = 4x - 12m + 4$ $-4(-x) = 4x$, and $-4(3m) = -12m$
                                              and $-4(-1) = +4$

$(y + 9)7 = 7y + 63$ $7 \cdot y = 7y$ and $7 \cdot 9 = 63$

$(x - 2y + 3)2 = 2x - 4y + 6$ $2 \cdot x = 2x$, and $2(-2y) = -4$,
                                              and $2(+3) = 6$

Distributive Law. Exercise 5. Use the Distributive Law to remove the parentheses.

1. $-2(x - 3) =$ 9. $-6(-3z - 2) =$

2. $-2(y + 9) =$ 10. $-1(-x-10) =$

3. $3(z - 1) =$ 11. $(-2x + 1)7 =$

4. $7(-2y + 3) =$ 12. $4(x + 3y + 1) =$

5. $4(-x - 2) =$ 13. $2(3y -5p - 4) =$

6. $(5x+2)3 =$ 14. $-5(a + 2b - 6) =$

7. $5(2m - 5) =$ 15. $(x - 2y + 7)3 =$

8. $-4(-2x + 3)$ 16. $(3x - 2) (-5) =$

Invisible Negative One

A negative sign in front of a number can always be replaced by -1.

•Examples: Reason

-5 is the same as $-1 \cdot 5$. $-1 \cdot 5 = -5$

-3 is the same as $-1 \cdot 3$. $-1 \cdot 3 = -3$

-x is the same as $-1x$. $-1x = -x$

The pattern in the above examples leads to the belief that:

$$-(x + 7) = -1(x + 7)$$

and

$$-(2x + 3y) = -1(2x + 3y).$$

It is useful to regard the -1 as already existing in expressions like *-3*, *-x* and *-(x+7)* , but the *-1* is invisible. It is made visible by replacing the negative sign by *-1* .

Distributive Law. Exercise 6. Make the invisible -1 visible by replacing the negative sign with -1.

1. -7	4. -(x + 9)	7. 5 - (x + 4)
2. -93	5. -(x - 6)	8. 7x - (3y + x)
3. -y	6. -(y + 2x - 3)	9. 4 - (x + 2) - (y - 2)

Distributing the Negative Sign.

The negative sign is distributed so that parentheses can be removed. Here are two methods to distribute the negative sign:

Method 1:

Step 1: Replace the negative sign by -1.
Step 2: Distribute the -1 to each term inside the parentheses.

•Example

-(-2x - 3y + 4)

$$= -1(-2x - 3y + 4) \quad \text{Step 1.}$$
$$= 2x + 3y - 4 \quad \text{Step 2. Since } -1(-2x) = 2x, \ -1(-3y) = +3y,$$
$$\text{and } -1(+4) = -4.$$

In the above example, each term in the answer is the same as each term in the parentheses except that all signs have changed. This observation leads to Method 2.

Method 2

Step 1: Change the sign of every term in the parentheses and drop the parentheses.

•Example:

-(3x - 2y - 5z + 4)	Remember 3x = +3x.
= -3x + 2y + 5z - 4	Step 1.

Distributive Law. Exercise 7. Distribute each negative sign.

1. -(-2x + 5)

2. -(+5y - 6)

3. -(7y - 4)

4. -(x + 2y - 6z - 9)

5. 7-(x + 2y) Recopy 7, then distribute the negative.

6. 5x - (-2y - 1) Recopy 5x first.

7. 14 - (2x - 6p + 4a - q)

8. -(x - y + 3) + 4z

Simplifying Expressions Containing Parentheses

To simplify expressions containing parentheses:
Step 1: Remove parentheses using the Distributive Law.
Step 2: Collect like terms.

•Example 1:

Simplify 2(x + 3) + 5x.

2(x + 3) + 5x	
2x + 6 + 5x	Step 1: Distribute the 2.
7x + 6	Step 2: Collect like terms.

•Example 2:

Simplify 2y - 2(4x + y - 3) .

2y - 2(4x + y - 3)	
2y - 8x - 2y + 6	Step 1: Distribute the -2.
-8x + 6	Step 2: Collect y terms: 2y - 2y = 0y = 0.

•Example 3:

Simplify 2x - (x + 3).

$$2x - (x + 3)$$
$$2x - 1(x + 3) \qquad \text{Invisible one.}$$
$$2x - x - 3 \qquad \text{Step 1: Distribute the -1.}$$
$$1x - 3 \qquad \text{Step 2: Collect like terms.}$$
$$x - 3$$

•Example 4:

Simplify 5y - (2x + 3y - 8x) + x.

$$5y - (2x + 3y - 8x) + x$$
$$5y - 2x - 3y + 8x + x \qquad \text{All terms in parentheses change signs.}$$
$$2y + 7x \qquad \text{Collect like terms: } 5 + (-3) = 2,$$
$$\text{and } -2 + 8 + 1 = 7.$$

Distributive Law. Exercise 8. Simplify.

1. $2(x + 1)$

2. $-3(y + 4)$

3. $5x + 2(x + 1)$

4. $-2(x + 4) + 2x$

5. $5(2y - 3) + 15$

6. $-3 + 5(p + 2)$

7. $-4(x + 2y) + y$

8. $-(x + 2) + 4x$

9. $x - (x + 2)$

10. $5 - (2y + 3)$

11. $6x - 2(3y + 3x)$ $\qquad$ Distribute -2 first.

12. $2z - 3(z + 1)$ $\qquad$ Distribute -3 first.

13. $10 + \frac{1}{2}(x + 6)$ $\qquad$ Hint: $\frac{1}{2} \cdot \frac{6}{1} = 3$

14. $\frac{1}{3}(6x + 1) + 3x$

15. $4x - \frac{1}{2}(8x + 1) + x$

16. $5(a + 1) + 2(a + 3 + b)$

Linear Equations

Evaluating an Expression

Evaluating an expression means finding the value of an expression when a variable is replaced by a number.

Step 1: Recopy the expression as it appears, except replace each variable with the given number and enclose it in parentheses.

Step 2: Follow the **Order of Operations** (see pre-algebra review).

•Example.

Evaluate $2y^2 - y + 5$, given that $y = -3$.

$2(-3)^2 - (-3) + 5$ Step 1: Even the negative in front of the y gets recopied.

$2 \cdot 9 - (-3) + 5$ Step 2: Exponent.

$18 - (-3) + 5$ Multiplication.

$18 + 3 + 5$ Subtracting a negative.

26 Addition

In other words, 26 is the value of $2y^2 - y + 5$ when y is -3.

Linear Equations. Exercise 1. Evaluate the following expressions.

1. $2y^2 - y + 5$ Given $y = -4$

2. $2y^2 - y + 5$ Given $y = 3$

3. $2x + 11$
    a. Given $x = -5$
    b. Given $x = 10$
    c. Given $x = 0$

4. $-5p + 1$
    a. Given $p = -2$
    b. Given $p = 1$

5. $-z + 7$
    a. Given $z = -2$
    b. Given $z = -1$
    c. Given $z = 7$

Linear Equations in One Variable and Solutions

$2x+11 = 5$, $y+4 = -3$, and $4z+0 = 5$ are examples of linear equations with one variable.

•More Examples:

Linear in x	Linear in y	Linear in m
-3x - 4 = 12	-y + 7 = 0	m - 4 = 8
x + 1 = 1	-29y + 29 = 29	-2m - 6 = 8

The **solution** to such equations is the number which makes the equation true. For instance, *-3* is the solution of $2x+11 = 5$ because replacing x with *-3* gives:

$$2x + 11 = 5$$
$$2(-3) + 11 = 5$$
$$-6 + 11 = 5 \text{ which is a true statement, since both sides equal 5.}$$

Therefore, *-3* is the solution to $2x+11 = 5$.

Any replacement of x which leads to a false statement is <u>not</u> the solution. For instance, *4* is not the solution to $2x+11 = 5$ because replacing x with *4* gives:

$$2x + 11 = 5$$
$$2(4) + 11 = 5$$
$$8 + 11 = 5$$
$$19 = 5 \text{ which is a false statement. Therefore, } 4 \text{ is not a solution.}$$

Linear Equations. Exercise 2. Determine the following by replacing the variable with the given number.

1. Is 1 the solution of -5x + 1 = -4 ?
2. Is 2 the solution of y + 4 = -3 ?
3. Is -7 the solution of y + 4 = -3 ?
4. Is 10 the solution of 9x + 5 = 5 ?
5. Is 0 the solution of 9x +5 = 5 ?
6. Is 7 the solution of -z + 7 = 0 ?
7. Is -3 the solution of -3x -4 = 12 ?
8. Is $\frac{5}{4}$ the solution of 4m + 0 = 5 ?

Solving Equations with the Addition Principle

The **solution** is the number which makes the equation true. An equation is said to be *solved* once the solution has been discovered. The process of finding the solution (also called *solving the equation*) often relies upon a concept known as the **Addition Principle**.

Addition Principle: It is legal to add or subtract the same number to or from both sides of any equation.

•Example:

---

1 + 4 = 5 is a true statement.

> Let's add any number, say 3, to both sides of the equation to get:
> $$1 + 4 + 3 = 5 + 3$$

This is also a true statement, since both sides now equal 8.

Using the Addition Principle to Isolate the Variable

To solve equations using the Addition Princple:

Step 1: Determine which number is on the same side of the equal sign as the variable.
Step 2: Add the opposite of that number to both sides of the equation.
Step 3: Collect like terms on each side. The variable is now alone (isolated) on one side of the equal sign. The number on the other side should be the desired solution.
Step 4: Check the solution in the original equation to be certain no mistakes have been made.

•Example 1:

Solve x - 10 = 4 .

Our goal is to get x alone (isolate the variable). To isolate *x* we need to get rid of *-10* . Add *10* to both sides to get:

$$x - 10 + 10 = 4 + 10$$
$$x + 0 = 14$$
$$x = 14$$

To check that 14 is the desired solution, replace *x* with *14* in the original equation:

$$x - 10 = 4$$
$$14 - 10 = 4 \text{ (replace x with 14)}$$
$$4 = 4 \text{ is true. Hence, 14 is the solution to } x - 10 = 4.$$

•Example 2:

Solve $21 = y + 15$.

Step 1: *15* is on the same side of the equal sign as *y* .

$$21 = y + 15$$
$$21 + (-15) = y + 15 + (-15) \qquad \text{Step 2: -15 is the opposite of 15.}$$
$$6 = y + 0$$
$$6 = y$$

Check by placing *6* in the original equation:

$$21 = y + 15$$
$$21 = 6 + 15$$
$$21 = 21 \text{ is true. Therefore, 6 is the solution to } 21 = y + 15.$$

•Example 3:

Solve $21 = y + 15$ (an alternate way).

To isolate *y* , get rid of *15* by subtracting *15* from both sides:
$$21 = y + 15$$
$$21 - 15 = y + 15 - 15$$
$$6 = y. \text{ Same as the solution in Example 2.}$$

Linear Equations. Exercise 3. Solve . Then check the solution.

1. $x - 3 = 4$ 5. $19 = y + 12$ 9. $x + 1.5 = 3.5$

2. $y + 2 = 1$ 6. $10 + x = -13$ 10. $z + \frac{1}{2} = \frac{5}{2}$

3. $7 = x - 4$ 7. $75 = m - 50$ 11. $-14 = 6 + x$

4. $z + 22 = 25$ 8. $10 = 5 + y$ 12. $-9 = -9 + y$

Solving Equations with the Multiplication Principle

The **solution** is the number which makes the equation true. An equation is said to be *solved* once the solution has been discovered. The process of finding the solution (also called *solving the equation*) often relies upon a concept known as the **Multiplication Principle.**

Multiplication Principle: It is legal to multiply or divide both sides of an equation by the same number.

•Example:

> $3 \cdot 5 = 15$ is a true statement.
>
> Let's multiply both sides by any number, say 2, to get $2 \cdot 3 \cdot 5 = 2 \cdot 15$. This is also a true statement, because both sides equal 30.

Using the Multiplication Principle

To solve equations using the multiplication principal:
Step 1: Determine the coefficient of the variable term.
Step 2: Isolate the variable by multiplying both sides by the reciprocal of that coefficient.
Step 3: Check the solution by inserting it into the original equation.

•Example 1:

> Solve $3x = 27$.
>
> **Step 1:** 3 is the coefficient of the variable term.
> **Step 2:** Multiply both sides by $\frac{1}{3}$ (the reciprocal of 3) to get:
>
> $$3x = 27$$
> $$\frac{1}{3}(3x) = \frac{1}{3}(27)$$
> $$\frac{1}{3} \cdot \frac{3}{1}x = \frac{1}{3} \cdot \frac{27}{1}$$
> $$\frac{3}{3}x = \frac{27}{3}$$
> $$1x = 9$$
> $$x = 9$$

To check that 9 is the desired solution, replace x with 9 in the original equation:

$$3x = 27$$

$$3(9) = 27$$

$$27 = 27 \text{ is true. Hence, 9 is the solution to } 3x = 27.$$

•Example 2.

Solve 3x = 27 (an alternative method)

Isolate x by dividing both sides by 3 to get:

$$3x = 27$$

$$\frac{3}{3}x = \frac{27}{3}$$

$$1x = 9$$

$$x = 9 \qquad \text{Same solution as Example 1.}$$

The next example summarizes both methods.

Example 3:

Solve 7 - 4y .

Step 1: -4 is the coefficient of the y-term. Our goal is to isolate y by eliminating - 4. There are two ways to get rid of -4:

Multiply by the Reciprocal OR Divide by -4
(The reciprocal of -4 is $-\frac{1}{4}$)

$$7 = -4y \qquad\qquad\qquad 7 = -4y$$

$$-\frac{1}{4}(7) = -\frac{1}{4}(-4y) \qquad\qquad \frac{7}{-4} = \frac{-4y}{-4}$$

$$-\frac{1}{4}\cdot\frac{7}{1} = -\frac{1}{4}\cdot-\frac{4}{1}y \qquad\qquad -\frac{7}{4} = \frac{\cancel{-4}y}{\cancel{-4}} \quad \text{Cancel -4s.}$$

$$-\frac{7}{4} = -\frac{1}{\cancel{4}}\cdot-\frac{\cancel{4}}{1}y \quad \text{4s cancel.} \qquad -\frac{7}{4} = y$$

$$-\frac{7}{4} = +y \qquad \text{Negatives cancel.}$$

•Example 4:

Solve -x = 5

Our goal is to get rid of the negative sign. Multiply both sides by -1 to get:

$$-x = 5$$
$$-1(-x) = -1 \cdot 5$$
$$x = 5$$

To check, replace x with -5 in the original equation:

$$-x = 5$$
$$-(-5) = 5 \quad \text{(replacing x with -5)}$$
$$5 = 5 \text{ is true. Hence, } x = -5 \text{ is a solution to } -x = 5.$$

Linear Equations, Exercise 4. Solve and check.

1. Solve by dividing both sides as in Example 2 and Example 3 (second way). Check each result.

   a. $5y = 10$ See Example 2   g. $-3y = -10$
   b. $-3x = 12$ See Example 3   h. $7x = 7$
   c. $5 = 2y$ i. $5y = 0$
   d. $-4x = 3$ j. $-3x = 3$
   e. $2 = 10y$ k. $-2x = -2$
   f. $8 = -2x$ l. $3 = -6y$

2. Solve a., b., h., i., and l. again. This time multiply by the reciprocal as in Example 1 and Example 3 (first way).

3. Solve as in Example 4:

   a. $-x = 1$ c. $4 = -x$
   b. $-y = -7$ d. $-y = 0$

4. Solve by multiplication using the reciprocal. Don't check.

   a. $\frac{1}{3}x = 4$ (Reciprocal is $\frac{3}{1}$)   c. $3 = \frac{1}{2}y$
   b. $\frac{3}{2}y = 3$ (Reciprocal is $\frac{2}{3}$)   d. $-\frac{5}{6}x = 20$ (Reciprocal is $\frac{-6}{5}$)

Solving Equations Using Both Addition and Multiplication Principles

In more involved equations, both the addition and multiplication principles are used together. To solve these equations:

Step 1: Use the distributive property to remove all parentheses;
Step 2: Combine like terms on each side;
Step 3: Use the addition principle (possibly many times) to get all the variable terms on one side and all non-variable terms on the other side;
Step 4: Use the multiplication principle to get the variable by itself;
Step 5: Check the solution.

•Example 1:

Solve $3x + 2 = 11$

There are no parentheses and no like terms on any one side, so skip Step 1 and Step 2.

$$3x + 2 = 11$$

Step 3: Addition principle. The 2 is a non-x term: move it to the other side.

$$3x + 2 - 2 = 11 - 2$$
$$3x + 0 = 9$$
$$3x = 9$$
$$\tfrac{1}{3}(3x) = \tfrac{1}{3} \cdot 9$$

Step 4: Multiplication principle.

$$\tfrac{1}{3} \cdot \tfrac{3}{1}x = \tfrac{1}{3} \cdot \tfrac{9}{1}$$
$$1x = \tfrac{9}{3}$$
$$x = 3$$

Check $x = 3$ in the original equation (Step 5).

$$3x + 2 = 11$$
$$3(3) + 2 = 11$$
$$9 + 2 = 11 \text{ is true.}$$

Hence, 3 is the solution to $3x + 2 = 11$.

•Example 2:

Solve and check $-3y + 2 = 1 - 4y$.

$$-3y + 2 = 1 - 4y$$

Step 3: By the addition principle, move the -4y to the other side.

$$-3y + 2 + 4y = 1 - 4y + 4y$$
$$y + 2 = 1$$

Collect terms.

$$y + 2 - 2 = 1 - 2$$

Step 3: Move the non-y term, 2, by subtracting it from both sides.

$$y = -1$$

Check y = -1 in the original equation.

$$-3y + 2 = 1 - 4y$$
$$-3(-1) + 2 = 1 - 4(-1)$$
$$3 + 2 = 1 + 4$$
$$5 = 5 \qquad \text{Hence, -1 is the solution.}$$

•Example 3:

Solve and check $2(x + 1) = 5x - (2x - 4)$.

$2(x + 1) = 5x - 1(2x - 4)$	The *invisible* 1 (page 10 - 11).
$2x + 2 = 5x - 2x + 4$	Step 1. Distribute the 2 and -1.
$2x + 2 = 3x + 4$	Step 2. Collect like terms.
$2x + 2 - 3x = 3x + 4 - 3x$	Step 3. Addition principle to to
$-1x + 2 = 4$	get x terms on the left side.
$-x + 2 - 2 = 4 - 2$	Step 3: Again to get all non-x
$-x = 2$	terms on the right side.
$(-1)(-x) = (-1)(2)$	Step 4. Multiplication principle.
$x = -2$	

Linear Equations. Exercise 5. Solve and check. Hints given below.

1. $5x - 7 = 3$ Steps 3 - 5 required.

2. $3y + 10 = 40$

3. $-11 = 6x + 7$

4. $3x + 4x = 49$ Steps 2, 4 and 5.

5. $10y - 7y = -33$

6. $3 = 12x - 8x$

7. $5x - 7 = 4x$ Steps 3 - 5.

8. $p + 2 = 17 - 4p$

9. $5(x + 2) = 20$ Most steps required.

10. $2(y - 1) + 1 = 3$

11. $4 - 3(2x + 1) = 7$ Distibute 3 with its negative.

12. $-(x - 2) = 5$ Remember $-(x - 2) = -1(x-2)$.

13. $-(2x + 1) = 3$ Distribute the negative sign.

14. $6(z - 1) = z + 4$

15. $4 - (2x - 3) = 2(2x + 1) + 5$

16. $3(2x + 3) = -2(x + 2) - 3$

Solving Equations Containing Fractions.

To solve equations containing fractions, we need a concept called the **least common multiple**.

The Least Common Multiple (LCM) of Two Numbers

The LCM is the smallest number that both numbers can divide into evenly. For example, the LCM of 6 and 9 is 18, since 18 is the smallest number that both 6 and 9 can divide into evenly.

To Solve Equations Containing Fractions

Step 1: If one side of the equal sign contains two or more terms, enclose these terms in parentheses.

Step 2: Find the LCM of the denominators. Multiply both sides of the equal sign by the LCM over 1.

Step 3: Use the Distributive Law to write the LCM next to each term inside the parentheses.

Step 4: Cancel denominators. Isolate variables as before, using the addition and multiplication principles.

•Example

Solve $\frac{1}{6}x - 2 = \frac{3}{4}$.

$(\frac{1}{6}x - 2) = \frac{3}{4}$ .... Step 1.

The LCM of 6 and 4 is 12. Multiply both sides of the equal sign by $\frac{12}{1}$.

$\frac{12}{1}(\frac{1}{6}x - 2) = \frac{12}{1} \cdot \frac{3}{4}$ .... Step 2.

$\frac{12}{1} \cdot \frac{1}{6}x - \frac{12}{1} \cdot 2 = \frac{12}{1} \cdot \frac{3}{4}$ .... Step 3.

$\frac{\overset{2}{\cancel{12}}}{1} \cdot \frac{1}{\cancel{6}}x - \frac{12}{1} \cdot 2 = \frac{\overset{3}{\cancel{12}}}{1} \cdot \frac{3}{\cancel{4}}$ .... Step 4.

$\frac{2}{1} \cdot \frac{1}{1}x - \frac{12}{1} \cdot \frac{2}{1} = \frac{3}{1} \cdot \frac{3}{1}$

$2x - 24 = 9$

$2x - 24 + 24 = 9 + 24$ .... Add 24 to both sides.

$2x = 33$

$\frac{2x}{2} = \frac{33}{2}$ .... Divide by 2.

$x = \frac{33}{2}$

Linear Equations. Exercise 6. In each of the following: a. Find the LCM of the denominator; b. solve, but don't check.

1. $\frac{1}{9}x + 2 = \frac{2}{3}$ LCM = 9.

2. $\frac{1}{5}x - 4 = \frac{1}{10}x$

3. $\frac{5}{6} = \frac{1}{4}x + 1$

4. $4(\frac{1}{4}x + \frac{1}{2}) = 4$ Use distributive property first

5. $3(\frac{1}{2}x + \frac{1}{6}) = 2$ Use distributive property first.

6. $-(x + \frac{1}{4}) = 7$ Distribute negative first.

Number Lines

Number Lines

The number line goes forever in the negative direction and forever in the positive direction.

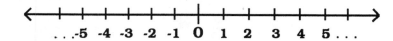

•Example:

Locate the following by making a dot above each position:

a. 4   b. -2   c. -5   d. $2\frac{1}{2}$ e. $-3\frac{1}{2}$ f. 0.

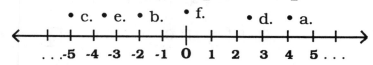

Number Lines. Exercise 1. Locate the following with dots:

a. 1   b. -5   c. 4.5   d. 6   e. -1   f. -3.5.

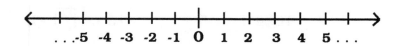

Inequalities

The symbols **<, >, ≤, ≥** are called inequality symbols. Inequality symbols are used to compare two numbers. For example:

8 > 5 means 8 is more than 5
1 < 4 means 1 is less than 4

The inequality symbol always points to the lesser of the two numbers. The lesser of the two numbers is the number farthest to the left on the number line.

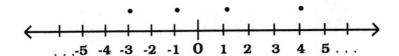

...-5 -4 -3 -2 -1 **0** 1 2 3 4 5...

•Example 1:

On the number line	We write.	We say.
1 is left of 4	1 < 4 or 4 > 1	One is less than four. or Four is more than one.
-3 is left of -1	-3 < -1 or -1 > -3	-3 is less than -1. or -1 is more than -3.

Number Lines. Exercise 2. Fill in the symbols (< or >).

1. 5 ☐ 10

2. 7 ☐ 3

3. -3 ☐ 2

4. -3 ☐ -2

5. -2 ☐ 7

6. 5 ☐ -100

7. -5 ☐ 100

8. -6 ☐ -4

9. 8 ☐ -2

10. 4 ☐ 5

11. -4 ☐ -5

12. -4 ☐ 5

13. $\frac{1}{2}$ ☐ 2

14. $\frac{1}{4}$ ☐ -$\frac{3}{4}$

15. -$\frac{1}{4}$ ☐ -1

≤, ≥ Symbols

≤ means the number to its left is less or equal to the number to its right.

≥ means the number to its left is more or equal to the number to its right.

•Examples:

2 ≤ 3 is true	5 ≥ 1 is true
2 ≤ 2 is true	5 ≥ 5 is true
2 ≤ 1 is false	5 ≥ 6 is false

Number Lines. Exercise 3. True or false.

1. $1 \leq 7$	4. $7 \geq 7$	7. $-3 \leq 5$
2. $1 \leq 1$	5. $-10 \leq -10$	8. $17 \geq 17$
3. $1 \leq 0$	6. $4 \geq 9$	9. $-4 \geq -3$

Solving Inequalities

$2x < 4$, $y - 5 \geq 7$, and $3x > -10 + x$ are examples of inequalities in one variable. To solve inequalities means to find all replacements of x that make the relationship true.

Addition Principle. It is legal to add or subtract any number to or from both sides of any inequality.

•Example:

Solve and graph $x - 5 \geq -7$.

Our goal is to isolate x. Add 5 to both sides.

$$x - 5 + 5 \geq -7 + 5$$
$x \geq -2$ This means that any replacement of x which is more than or equal to -2 makes the original inequality true.

The graph of $x \geq -2$ is

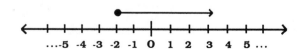

The ray (the line extending from the dot) represents all solutions that are more than -2. The solid dot shows that -2 is also a solution. If we had $x > -2$ instead of $x \geq -2$, the dot would be an open dot:

The graph of $x > -2$ is:

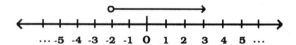

Number Lines. Exercise 4.

1. In the above example: the solution to $x - 5 \geq -7$ was found to be $x \geq -2$
   (That is, any number more than or equal to -2 will make $x - 5 \geq -7$ true).

a. 6 is more than -2. In x - 5 ≥ -7, replace x by 6 to show that 6 makes x - 5 ≥ -7 true.

b. -3 is not more than -2. In x - 5 ≥ -7 , replace x by -3 to show that -3 makes x - 5 ≥ -7 false.

2. Solve by the addition method, then graph.

a. x - 2 ≥ 1   e. p -7 ≤ -7
b. x + 3 < 5   f. 3x ≥ 2x - 4 (Subtract 2x from both sides)
c. x - 4 ≤ -3   g. 6y < 5y + 1
d. x + 17 > 15   h. 12z > 11z

Multiplication Principle. It is legal to multiply or divide both sides of an inequality by the same number. However, when multiplying or dividing by a negative number the inequality symbol must flip directions.

•Example 1:

---

2 < 4.

2 < 4 is a true inequality. Let's see what happens when both sides are multiplied by +3 and then by -3.

Multiply both sides by 3.   Multiply both sides by -3.

   2 < 4   2 < 4
3(2) $\boxed{?}$ 3(4)   -3(2) $\boxed{?}$ -3(4)
   6 < 12   -6 > -12

Inequality symbol unchanged.   Symbol must switch to get a true statement.

•Example 2.

Solve and graph -5x < 20.

-5x < 20   Divide by -5 and switch inequality

$\frac{-5}{-5}$ x > $\frac{20}{-5}$

   x > -4   ←|—|—|—|—|—|—|—|—|—|—|—|→
                        ... -5 -4 -3 -2 -1 0 1 2 3 4 5 ...

•Example 3:

Solve and graph 4y ≥ -8.

$\frac{4y}{4} \geq \frac{-8}{4}$

Divide by +4 and do not switch.

y ≥ -2

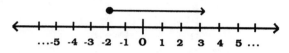

....-5 -4 -3 -2 -1 0 1 2 3 4 5 ...

•Example 4.

Solve and graph $-\frac{1}{3}x + 5 \geq 4$ using addition and multiplication together.

$-\frac{1}{3}x + 5 \geq 4$

$-\frac{1}{3}x + 5 - 5 \geq 4 - 5$ Addition principle.

$-\frac{1}{3}x \geq -1$

$-\frac{3}{1}(-\frac{1}{3}x) \leq (-\frac{3}{1})(-1)$ Multiply by $-\frac{3}{1}$ and switch.

$x \leq 3$

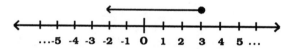

....-5 -4 -3 -2 -1 0 1 2 3 4 5 ...

Number Lines. Exercise 5.   Solve, but don't graph.

1. -3x ≤ 15

2. 3x < 21 Divide by +3, don't switch

3. -2y > -4

4. 2x > -4

5. 20p ≥ 10

6. $-\frac{1}{2}x \geq 3$

7. -7x ≥ 3

8. 2x -1 ≤ 5 Add 1 first

9. -3y + 2 ≥ -1

10. 5x - 8 < 2x + 1 Subtract 2x both sides

11. 2x + 6 ≤ 4x + 1

12. a - 3 > 5a -10

Review

Chapter 2

1. List the terms, coefficients, and like terms:

   $-3x + 2y - x + 19 - 4$

2. In problem 1, collect like terms.

3. Distribute the number or negative sign:

   a. $-4(3y - 2x + 1)$ c. $-1(x - 2)$

   b. $(6p - 1)\, 7$ d. $-(y - 4x + 1 - p)$

4. Simplify:

   a. $17 + 2(x - 5)$ c. $-8a - 4(3 - 2a)$ e. $4 - (-y + 7)$

   b. $-4(p - 2) + p$ d. $-(3a - 6) + 5a$ f. $-(x - 2y + 3x) + 2(3x + y)$

Chapter 3

1. a. Is 9 a solution to $5x - 40 = 5$

   b. Is -4 a solution to $x - 4 = -8$?

   c. Is 0 a solution to $5x - 5 = 0$?

   d. Is 0 a solution to $25x = 0$?

2. Solve:

   a. $14 = 3 + y$ e. $4x = 20$ i. $6 = -y$

   b. $x - 4 = 0$ f. $-7x = 14$ j. $0 = 10x$

   c. $-3 + p = 2$ g. $4x = 3$ k. $\frac{1}{2}x = 9$

   d. $a + 5 = 5$ h. $-7 = -5y$ l. $-\frac{3}{2}y = \frac{1}{4}$

3. Solve:

       a. -x + 7 = 10   e. 2y + 1 = y + 10   i. -9(x - 2) = 0

       b. -4 - y = 1   f. 5p + 4 = 2p - 5   j. 5m = 2 - 3(m - 2)

       c. 2x + 1 = 6   g. 4x = 3x   k. -(a - 3) = 4

       d. 5x + 2 = -8   h. 2(5y - 3) = 4   l. 4x - (x - 1) = 2(x + 3)

4. Solve.

       a. $\frac{3}{2}x = 4$ LCM = 2

       b. $8(\frac{1}{2} + \frac{x}{4}) = 1$ Distribute first.

       c. $\frac{1}{6}x - 2 = \frac{2}{3}$ LCM = 6

Chapter 4

1. Fill in < or > :

       a. -5 ☐ 0   b. -7 ☐ -3   c. -2 ☐ 1

2. a. Is 2 a solution to 5x + 1 ≥ 4?
   b. Is 3 a solution to 4x-10 ≤ 2 ?
   c. Is 3 a solution to -x + 6 ≥ 6?

3. Solve and graph:

       a. y + 4 > 3   c. -3x ≤ 6

       b. 5p ≤ -10   d. 2x -1 < 5

Properties of Exponents

Factors and Fractions

Numbers connected to each other by multiplication are called **factors**. For instance in $2 \cdot 7 = 14$, the factors are 2 and 7.

When the same factor appears in both the numerator and the denominator of a fraction, it can be crossed out and replaced by the number 1.

Example:

$$\text{Consider the fraction } \frac{5 \cdot 2}{2} .$$

$$\frac{5 \cdot 2}{2} = \frac{5 \cdot \overset{1}{\cancel{2}}}{\cancel{2}_1} = \frac{5 \cdot 1}{1} = \frac{5}{1} = 5$$

The same principle applies to variable factors.

Example:

$$\text{Consider } \frac{y}{4 \cdot y} .$$

$$\frac{y}{4 \cdot y} = \frac{\overset{1}{\cancel{y}}}{4 \cdot \cancel{y}_1} = \frac{1}{4 \cdot 1} = \frac{1}{4}$$

More examples:

$$\frac{x \cdot x}{x} = \frac{\overset{1}{\cancel{x}} \cdot x}{\cancel{x}_1} = \frac{1 \cdot x}{1} = \frac{x}{1} = x$$

$$\frac{5 \cdot 5 \cdot 2}{5 \cdot 5 \cdot 3} = \frac{\overset{1}{\cancel{5}} \cdot \overset{1}{\cancel{5}} \cdot 2}{\cancel{5}_1 \cdot \cancel{5}_1 \cdot 3} = \frac{1 \cdot 1 \cdot 2}{1 \cdot 1 \cdot 3} = \frac{2}{3}$$

$$\frac{a \cdot a}{a \cdot a \cdot a} = \frac{\overset{1}{\cancel{x}} \cdot \overset{1}{\cancel{x}}}{\cancel{x}_1 \cdot \cancel{x}_1 \cdot a} = \frac{1}{a}$$

Properties of Exponents, Exercise 1. Simplify each expression by crossing out factors and replacing them with a 1.

1. $\dfrac{2 \cdot 7}{7}$

4. $\dfrac{2 \cdot 3}{2 \cdot 3 \cdot x}$

7. $\dfrac{x \cdot x \cdot x}{x \cdot x \cdot x}$

2. $\dfrac{3 \cdot 2}{3 \cdot 5 \cdot 2}$

5. $\dfrac{a \cdot a \cdot a \cdot a}{a}$

8. $\dfrac{b \cdot b \cdot b \cdot b}{b \cdot b \cdot b}$

3. $\dfrac{4 \cdot y}{4}$

6. $\dfrac{y \cdot y}{y \cdot y \cdot y}$

9. $\dfrac{y \cdot y}{y \cdot y \cdot y \cdot y}$

Exponential Expressions

In the expression 5^4,   5 is called the **base** and ^{4} is called the **exponent**. 5^4 is read "5 raised to the 4th power".

$$5^4 = 5 \cdot 5 \cdot 5 \cdot 5 \qquad\qquad \text{4 factors of 5}$$

More examples:

$3^5 = 3 \cdot 3 \cdot 3 \cdot 3 \cdot 3$	5 factors of 3.
$\left(\frac{4}{7}\right)^2 = \frac{4}{7} \cdot \frac{4}{7}$	2 factors of $\frac{4}{7}$.
$(-2)^4 = (-2)\,(-2)\,(-2)\,(-2)$	4 factors of -2.
$x^3 = x \cdot x \cdot x$ or xxx	3 factors of x.
$x^4 = xxxx$	4 factors of x.
$x^1 = x$	1 factor of x.
$y^2 = yy$	2 factors of y.
$(7x)^2 = (7x)\,(7x)$	2 factors of 7x.
$(y^2)^3 = (y^2)\,(y^2)\,(y^2)$	3 factors of y^2.

The Invisible One Exponent

If no exponent is shown, the exponent is assumed to be **1** . The 1 exponent is invisible.

•Examples:

5 means 5^1	because $5^1 = 5.$	1 factor of 5
x means x^1	because $x^1 = x.$	1 factor of x
y means y^1	because $y^1 = y.$	1 factor of y

Properties of Exponents, Exercise 2. Write each as a product of factors.

1. 4^3 3. $(-4)^5$ 5. y^7 7. z^3 9. $(3b)^4$

2. $(\frac{1}{3})^2$ 4. x^6 6. y^1 8. $(2y)^3$ 10. $(x^2)^3$

Write the following using exponents.

11. $3 \cdot 3 \cdot 3$ 14. $x\,x\,x$ 17. $p\,p\,p\,p\,p$

12. $4 \cdot 4$ 15. 7 18. $(2a)\,(2a)\,(2a)$

13. $x\,x\,x\,x\,x$ 16. y 19. $(x^2)\,(x^2)\,(x^2)$

Product Rule

To multiply expressions with the same base, use the Product Rule. 4^2 and 4^3 are examples of expressions with the same base.

Product Rule: Recopy the base without changing it, then add the exponents together.   In other words $x^m \cdot x^n = x^{m+n}$.

•Examples:

$$4^2 \cdot 4^3 = 4^{2+3} = 4^5$$

$$x^8 \cdot x^2 = x^{8+2} = x^{10}$$

$$x^5 \cdot x^2 = x^{5+2} = x^7$$

In general $x^m \cdot x^n = x^{m+n}$.

If you care to see why the Product Rule works, consider the first example again by a longer route:

$$
\begin{aligned}
4^2 \cdot 4^3 &= (4 \cdot 4) \cdot (4 \cdot 4 \cdot 4) &&\text{Since } 4^2 = 4 \cdot 4 \text{ and } 4^3 = 4 \cdot 4 \cdot 4 \\
&= 4 \cdot 4 \cdot 4 \cdot 4 \cdot 4 &&\text{Drop the parentheses.} \\
&= 4^5 &&\text{5 factors of 4.}
\end{aligned}
$$

The base is unchanged. The final quantity of 4s is the same as the powers added together. That is, $5 = 2 + 3$.

Power Rule

To raise a power to a power, use the Power Rule.

Power Rule: Recopy the base, then multiply the powers together. In other words: $(x^m)^n = x^{m \cdot n}$.

•Examples:

$$
\begin{aligned}
(5^7)^4 &= 5^{7 \cdot 4} = 5^{28} \\
(x^2)^3 &= x^{2 \cdot 3} = x^6 \\
(x^5)^2 &= x^{5 \cdot 2} = x^{10}
\end{aligned}
$$

In general, $(x^m)^n = x^{m \cdot n}$.

If you care to see why the Power Rule works, consider example 2 again by a longer route:

$$
\begin{aligned}
(x^2)^3 &= (x^2)(x^2)(x^2) && \text{3 factors of } x^2. \\
&= (x\,x)(x\,x)(x\,x) && \text{Each } x^2 \text{ is } x\,x. \\
&= x\,x\,x\,x\,x\,x && \text{Dropping the parentheses.} \\
&= x^6 && \text{6 factors of } x.
\end{aligned}
$$

The final number of xs is the same as the powers multiplied. That is, $6 = 2 \cdot 3$.

Quotient Rule

To divide expressions having the same base, use the Quotient Rule.

Quotient Rule: Recopy the base, then subtract the exponents. In other words: $\dfrac{x^m}{x^n} = x^{m-n}$

•Examples:

1. $\dfrac{7^{10}}{7^4} = 7^{10-4} = 7^6$

2. $\dfrac{x^4}{x^3} = x^{4-3} = x^1 = x$

3. $\dfrac{y^5}{y^2} = y^{5-2} = y^3$

In general: $\dfrac{x^m}{x^n} = x^{m \cdot n}$

If you care to see why the Quotient Rule works, consider Example 3 by a longer route:

$$\frac{y^5}{y^2} = \frac{y\,y\,y\,y\,y}{y\,y} = \frac{\cancel{y}\,\cancel{y}\,y\,y\,y}{\cancel{y}\,\cancel{y}} = y\,y\,y = y^3$$

The final number of y s is the same as the power of the numerator minus the power of the denominator. That is, 3 = 5 - 2.

Properties of Exponents, Exercise 3. Simplify these expressions.

1. $x^5 \cdot x^2$

2. $(x^5)^2$

3. $\dfrac{x^5}{x^2}$

4. $x^7 \cdot x^3$

5. $(x^7)^3$

6. $(x^3)^7$

7. $\dfrac{x^7}{x^3}$

8. $(y^{10})^4$

9. $y^4 \cdot y^{10}$

10. $2^4 \cdot 2^3$

11. $(2^3)^4$

12. $\dfrac{2^4}{2^3}$

13. $x^7 \cdot x$

14. $6^8 \cdot 6$

15. $\dfrac{x^8}{x}$

16. $4 \cdot 4^2$

17. $(y^2)^1$

18. $\dfrac{y^2}{y}$

19. $5^9 \cdot 5^7$

20. $\dfrac{5^9}{5^7}$

21. $\dfrac{5^9}{5}$

22. $y \cdot y$ Hint: $y^1 \cdot y^1$

23. $x \cdot x$

24. $\dfrac{x}{x}$ Hint: $\dfrac{x^1}{x^1}$

Parentheses Raised to a Power

Parentheses Rule: Raise every factor in the parentheses to the power. In other words: $(abc)^m = a^m\, b^m\, c^m$. Then use the Power Rule when needed.

Examples:

$$(2x)^3 = 2^3 \cdot x^3 = 8x^3$$
$$(5y^4)^2 = 5^2\,(y^4)^2 = 25y^8 \qquad \text{By the Power Rule}$$
$$(4^2\,bc)^3 = (4^2)^3\,b^3\,c^3 = 4^6\,b^3\,c^3 \qquad \text{By the Power Rule}$$

In general : $(abc)^m = a^m\, b^m\, c^m$.

If you care to understand why the Parentheses Rule works, consider the first example by a longer route:

$$(2x)^3 = (2x)(2x)(2x) \qquad \text{3 factors of 2x.}$$
$$= 2 \cdot x \cdot 2 \cdot x \cdot 2 \cdot x \qquad \text{All factors are multiplied.}$$
$$= 2 \cdot 2 \cdot 2 \cdot x \cdot x \cdot x \qquad \text{Change order by Commutative Law.}$$
$$= 2^3 x^3 \qquad \text{Both factors appear 3 times.}$$

The power on each factor is the same as the power outside the parentheses.

Fractions: Raise both the numerator and the denominator to the power.

In other words: $\left(\dfrac{a}{b}\right)^m = \dfrac{a^m}{b^m}$. Then use the Power Rule when it is needed.

•Examples:

1. $\left(\dfrac{5}{x}\right)^2 = \dfrac{5^2}{x^2} = \dfrac{25}{x^2}$

2. $\left(\dfrac{y^2}{2}\right)^3 = \dfrac{(y^2)^3}{2^3} = \dfrac{y^6}{8}$

3. $\left(\dfrac{x}{3^2 y}\right)^4 = \dfrac{x^4}{(3^2 y)^4} = \dfrac{x^4}{3^8 y^4}$

If you care to see why this procedure works, consider example 1 again by a longer route:

$$\left(\dfrac{5}{x}\right)^2 = \dfrac{5}{x} \cdot \dfrac{5}{x} \qquad \text{2 factors of } \dfrac{5}{x}$$

$$= \dfrac{5 \cdot 5}{x \cdot x} = \dfrac{5^2}{x^2}$$

Properties of Exponents, Exercise 4. Simplify these expressions.

1. $(xy)^2$

2. $(2y)^3$

3. $(4y^3)^2$

4. $(4^3 y)^5$

5. $(3^4 ab)^2$

6. $(a\, x^4 y^2)^4$

7. $\left(\dfrac{2}{3}\right)^2$

8. $\left(\dfrac{x}{3}\right)^3$

9. $\left(\dfrac{4}{y^3}\right)^3$

10. $\left(\dfrac{7}{xy^5}\right)^2$

11. $\left(\dfrac{7}{xy^3}\right)^2$

12. $\left(\dfrac{5a^3}{b^3}\right)^3$

Negative Exponents

2^{-3} and 5^{-2} are examples of expressions with negative exponents. Such expressions are rewritten as fractions using the Negative Exponent Rule.

Negative Exponent Rule: Rewrite the expression as a fraction. Place a 1 in the numerator. In the denominator, recopy the base and change the power to its opposite. In other words: $x^{-m} = \frac{1}{x^m}$.

Examples:

1. $2^{-3} = \frac{1}{2^3}$ or $\frac{1}{8}$

2. $5^{-2} = \frac{1}{5^2}$ or $\frac{1}{25}$

3. $x^{-4} = \frac{1}{x^4}$

In general: $x^{-m} = \frac{1}{x^m}$.

If you wish to see why the Negative Exponent Rule works, consider the fraction $\frac{2^2}{2^5}$:

$$\frac{2^2}{2^5} = \frac{2 \cdot 2}{2 \cdot 2 \cdot 2 \cdot 2 \cdot 2} = \frac{\cancel{2} \cdot \cancel{2}}{\cancel{2} \cdot \cancel{2} \cdot 2 \cdot 2 \cdot 2} = \frac{1}{2 \cdot 2 \cdot 2} = \frac{1}{2^3}$$

On the other hand the Quotient Rule says that:

$$\frac{2^2}{2^5} = 2^{2-5} = 2^{-3}$$

Therefore $2^{-3} = \frac{1}{2^3}$, because both sides are equal to $\frac{2^2}{2^5}$.

Properties of Exponents, Exercise 5. Use the Negative Exponent Rule to rewrite each as a positive exponent.

1. $x^{-7} = \frac{1}{x^7}$ 5. x^{-73} 9. x^{-1}

2. x^{-5} 6. 2^{-4} 10. y^{-1}

3. y^{-2} 7. 3^{-3} 11. 5^{-1}

4. y^{-12} 8. 5^{-2} 12. $(3x)^{-1}$

Zero Exponent Rule

Zero Exponent Rule: Any base (other than 0) raised to the 0 power is always equal to 1. In other words: $x^0 = 1$, whenever $x \neq 0$.

•Examples:

1. $4^0 = 1$
2. $(117)^0 = 1$
3. $y^0 = 1$
4. $(349x^2 y^4)^0 = 1$ If the variables are not zero.

If you care to see why the Zero Exponent Rule works, consider the fraction: $\dfrac{4^2}{4^2}$.

$$\frac{4^2}{4^2} = \frac{16}{16} = 1$$

On the other hand, the Quotient Rule says, that $\dfrac{4^2}{4^2} = 4^{2-2} = 4^0$.

Hence, $4^0 = 1$, because both sides equal $\dfrac{4^2}{4^2}$.

Properties of Exponents, Exercise 6. Simplify using the Zero Exponent Rule.

1. $(17)^0$ 4. $(xyz)^0$

2. $(4,986)^0$ 5. $\left(\dfrac{49x^4y^3}{3a^3b^7}\right)^0$

3. x^0

Summary of Exponent Rules

Listed below are all of the Exponent Rules discussed so far.

1. $x^m \cdot x^n = x^{m+n}$ Product Rule.

2. $(x^m)^n = x^{m \cdot n}$ Power Rule.

3. $\dfrac{x^m}{x^n} = x^{m-n}$ Quotient Rule.

4. $(abc)^m = a^m b^m c^m$ Parentheses Rule.

5. $\left(\dfrac{a}{b}\right)^m = \dfrac{a^m}{b^m}$ Fraction Rule.

6. $x^{-m} = \dfrac{1}{x^m}$ Negative Exponent Rule.

7. $x^0 = 1 \quad x \neq 0$ Zero Exponent Rule.

Exponent Rules Applied to Negative and Zero Exponents

The first five rules of the previous summary can apply even when the exponents are negative or zero.

•Examples

1. $x^{-7} \cdot x^4 = x^{-7+4} = x^{-3}$ By the Product Rule

               $= \dfrac{1}{x^3}$ By the Negative Exponent Rule.

2. $(y^{-3})^2 = y^{-3 \cdot 2} = y^{-6}$ By the Power Rule.

               $= \dfrac{1}{y^6}$ By the Negative Exponent Rule.

3. $\left(\dfrac{4^0}{4^7}\right) = 4^{0-7} = 4^{-7}$ By the Quotient Rule.

               $= \dfrac{1}{4^7}$ By the Negative Exponent Rule.

4. $(x^{-2}y)^{-3} = (x^{-2})^{-3} y^{-3}$ By the Parentheses Rule.

         $= x^6 y^{-3}$ By the Power Rule.

         $= x^6 \cdot \dfrac{1}{y^3}$ By the Negative Exponent Rule.

•More examples:

1. $x^{-7} \cdot x^{-4} = x^{-7+(-4)} = x^{-11}$ By the Product Rule.

                 $= \dfrac{1}{x^{11}}$

2. $\dfrac{x^5}{x^{-3}} = x^{5-(-3)}$ By Quotient Rule.

         $= x^{5+3} = x^8$

3. $\left(\dfrac{4^{-2}}{4^{-6}}\right) = 4^{-2-(-6)}$ By the Quotient Rule.

         $= 4^{-2+6}$

         $= 4^4$

Properties of Exponents, Exercise 7.   Simplify these expressions.

1. $x^7 \cdot x^5$

2. $x^7 \cdot x^{-3}$

3. $x^{-7} \cdot x^3$

4. $x^{-7} \cdot x^{-3}$

5. $\dfrac{x^6}{x^4}$

6. $\dfrac{x^6}{x^{-4}}$

7. $\dfrac{x^{-6}}{x^4}$

8. $\dfrac{x^{-6}}{x^{-4}}$

9. $(y^3)^{-2}$

10. $(y^{-4})^{-3}$

11. $3^2 \cdot 3^{-5}$

12. $\dfrac{x^{-5}}{x^{-3}}$

13. $\dfrac{x}{x^3}$

14. $\dfrac{x}{x^2}$

15. $\dfrac{y}{y^{-3}}$

16. $(x^{-1}y^2)^{-2}$

17. $(3^{-2} y^{-3})^{-4}$

18. $\left(\dfrac{x^{-1}}{y^{-3}}\right)^{-2}$

19. $(9x^{-11})^0$

20. $\dfrac{x^0}{x^{-4}}$

Polynominals

Degree and Order

$8y^3 - x^2y^5 + 6$ is an example of a polynomial.

When working with polynomials, the words **terms** and **coefficients** have the same meaning as in Chapter 2.. For example, in $8y^3 - x^2y^5 + 6$ the terms are $8y^3$, $- x^2y^5$, and 6. The coefficients are 8, -1, and 6.

The **degree of a term** refers to the total number of variable factors in the term. In the above example, the degree of $8y^3$ is 3, since $8y^3$ contains three factors of y. The degree of $- x^2y^5$ is 7, since $- x^2y^5$ contains 2 factors of x and 5 factors of y. The degree of 6 is 0, since the term 6 has no variables at all.

In general, the degree of a term is found by adding its variable powers together.

The **degree of a polynomial** is identical to the degree of its highest term. For example, $8y^3 - x^2y^5 + 6$ is a 7th degree polynomial, since its highest degree term, $- x^2y^5$, has a degree of 7.

• Another example:

> In $6y - 2x^4y^5 + x^3y$ the terms are $6y$, $-2x^4y^5$, and x^3y. The degree of $6y$ is 1. The degree of $-2x^4y^5$ is 9. The degree of x^3y is 4. The highest degree term is $-2x^4y^5$. The degree of the polynomial is 9.

A polynomial is in **descending order** if the powers on the variable decrease from left to right.

• Examples:

> | | |
> |---|---|
> | $y^5 + y^3 + y + 1$ | is in descending order. |
> | $7x^{10} + 5x^9 + -3$ | is in descending order. |
> | $4x^3 + x^{10} - 5x$ | is **not** in descending order. |

Polynomials, Exercise 1. For the following, find the degree of each term and the degree of the polynomial.

1. $3y^2 - y + 4$
2. $2x^5 - x$
3. $x^2y^5 + 2x^7y$

4. $x^2 + y^4 + 9$
5. $4a^2 b^3 + 8 a^1 b^4$
6. $x^2y^3z^5 + 8$

Rearrange each in descending order.

7. $x^2 + x^{10} = x^{10} + x^2$
8. $x^3 + x^9 =$
9. $3 + y^2 + 4y^4 =$
10. $5 + 8y^8 + 6y =$
11. $x + 3 + x^2 + x^3 =$

Types of Polynomials

A polynomial that has three terms is called a **trinomial**. A polynomial that has two terms is called a **binomial**. A polynomial that has only one term is called a **monomial**. For instance, $8y^3 - x^2y^5 + 6$ is a trinomial; $2x^5 - x$ is a binomial; and, $21y^{10}$ is a monomial. A polynomial that has more than three terms has no special name.

•More examples:

Monomial	Binomial	Trinomial
10	$x^4 + 3$	$x^4 - 2x^2 + 3$
y^2	$x - y^2$	$2y + 4 - 6y^2$
$2xy$	$7xy + y$	$x^2y^2 + 3x + 10$

An expression is not a polynomial if the variable is raised to a negative power. For example, $2x^{-3} + y^{-2}$ is not a polynomial.

Polynomials, Exercise 2. For each of the following: (1) State whether it is a polynomial and (2) If it is, state whether it is a monomial, binomial, or trinomial.

1. $2x^4 - 3$
2. $7x^2 + 3y^{-4}$
3. $4x^{11}$
4. x^{-2}
5. $x^2 - 3 + x^{10}$
6. 4
7. $x^2 + x^4 + 2y^7 - 3xy$
8. $x^2 - y + 7$

Polynomials with Like Terms

Like terms are terms containing the same letters raised to the same powers.

•Example

Like Terms	**Not** Like Terms
$4y^3$ and $3y^3$	x^2 and x (power difference)
xy and $2xy$	$7x$ and $7z$ (letter difference)
$7x^3y$ and $4x^3y$	$3xy^2$ and x^2y (power difference)

To Collect Like Terms: Add the coefficients of the like terms together, followed by their common letters and powers.

•Example: Reason

1. $3x^5 + 7x^5 = 10x^5$	$3 + 7 = 10$.
2. $6x^2 + 5x + 2x^2 - x$ $= 8x^2 + 4x$	$6 + 2 = 8$ and $5 - 1 = 4$.
3. $4a^7b - 9a^7b + 2ab^3 - 2ab^3$ $= -5a^7b$	$4 + (-9) = -5$ and $2 - 2 = 0$.
4. $3x^7y + 5xy^7 + 6x^7y$ $= 9x^7y + 5xy^7$	$5xy^7$ is not a like term.

Polynomials, Exercise 3.

1. $2x^2 + 5x^2$

2. $4x^5 - x^5$

3. $4y^7 + 3x + 2y^7 - 5x$

4. $2x^3 + 3x + 5x^3 + x$

5. $5y^2 - 2y + 8y + 7$

6. $10x^2 - 4x - 12x^2 - 15x$

7. $11x^6y - 5x^6y + 3 + 2$

8. $xy^5 + 4x + 3xy^5$

9. $y^7 - 3y^4 + 2y$

10. $a^3b + 2ab - 2ab$

11. $2xy^5 + 7xy^5 + 2x^5y$

12. $8 - 2 + a^2 - 3b - a^2$

Adding and Subtracting Polynomials

To Add Polynomials: Recopy without the parentheses and collect like terms.

•Example:

> Add $(4x^3y + 5) + (5 - 7x^3y)$.
>
> $4x^3y + 5 + 5 - 7x^3y$ Drop parentheses.
> $-3x^3y + 10$ Collect like terms.

To Subtract Polynomials:
    Step 1: Remove the parentheses. Remember that negative signs
               distribute to each term in the second parentheses.
    Step 2: Collect terms.

•Example

> Subtract $(x^4 + 2x) - (x^4 - 2x + 5)$
>
> $x^4 + 2x - x^4 + 2x - 5$ Remove parentheses, change signs.
> $= 4x - 5$ Step 2: $x^4 - x^4 = 0$ $2 + 2 = 4$.

Polynomials, Exercise 4. Add.

1. $(4x + 7) + (3x - 5)$
2. $(12x^7 - 9y) + (3x^7 + y)$
3. $(5y^2 - 8y) + (2y^2 + 11y)$
4. $(x^2 - x + 1) + (x - 1)$
5. $(y^3 - xy) + (y^3 + xy)$
6. $(2xy + 3x) + (x - 6xy)$

Subtract

7. $(5x + 3) - (2x - 8)$
8. $(4x^{15} - 2y) - (x^{15} + 3y)$
9. $(4a^3 - 2b^5) - (4a^3 - 2b^5)$
10. $7xy^2 - (5xy^2 - 3)$
11. $(y^2 + 3y) - (2y^2 + 6y)$
12. $9x^2y - (7x + 3 - 2x^2y)$

Multiplying Polynomials Together

To Multiply a Monomial to a Monomial

Multiply coefficients together and use the Product rule to multiply like bases.

•Example:

Multiply $(2x^9y^3)$ $(-7x^4y)$

$\quad\quad\quad = -14x^{13}y^4$ $\quad\quad$ Since $2 \cdot {}^-7 = -14$; $x^9 \cdot x^4 = x^{13}$; and $y^3 \cdot y = y^4$.

If you wish to see why this process works, consider the example again by a longer route:

$$(2x^9y^3) \; (-7x^4y) \quad = 2 \cdot x^9 \cdot y^3 \cdot -7 \cdot x^4 \cdot y$$
$$= 2 \cdot -7 \cdot x^9 \cdot x^4 \cdot y^3 \cdot y \quad\quad \text{Commutative Property.}$$
$$= -14x^{13}y^4$$

Polynomials, Exercise 5.   Multiply.

1. $(5x^2)$ $(3x^4)$ 6. $(2x)$ (x) 11. (y^4) $(-xy)$

2. $(7y^9)$ $(-10y^3)$ 7. (y) $(3y)$ 12. (xy) (zy)

3. $(-2x^3)$ $(-6x)$ 8. (a) (a) 13. (pq^2) $(2p^2q)$

4. $(3x^4y)$ $(2x^3y^2)$ 9. $(-3x^{20}y^{20})$ $(-3x^{20}y^{20})$ 14. (-4) $(2x^2)$

5. $(-5x^3y^2)$ (x^5y) 10. $(-7y)$ $(2x^2)$

To Multiply a Polynomial by a Monomial

Distribute the monomial to every term of the polynomial.

•Example 1:

In $5x^2(x^3 - 4x)$ the $5x^2$ multiplies both x^3 and $-4x$:

$\quad\quad 5x^2 \cdot x^3 = 5x^5$

$\quad\quad 5x^2 \; (x^3 - 4x) \quad = \quad 5x^5 \quad - 20x^3$

$\quad\quad\quad 5x^2 \cdot -4x = -20x^3$

The entire process without the arrows looks like this:

$$5x^2(x^3 - 4x) = 5x^5 - 20x^3$$

•Example 2:

$3xy(x^2 + 2xy - 4)$	Distribute $3xy$.
$= 3x^3y + 6x^2y^2 - 12xy$	Because: $3xy \cdot x^2 = 3x^3y$, and $3xy \cdot 2xy = 6x^2y^2$, and $3xy(-4) = -12xy$.

•More examples: Reasons

$7(2x^2 - 3y) = 14x^2 - 21y$

$x(2x + y) = 2x^2 + xy$ $x \cdot 2x = 2x^2$ and $x \cdot y = xy$

$-6y(3y - 2y^3) = -18y^2 + 12y^4$ $-6y \cdot 3y = -18y^2$ and $-6y \cdot -2y^3 = 12y^4$

$2xy^2(y^3 - 4) = 2xy^5 - 8xy^2$ $2xy^2 \cdot y^3 = 2xy^5$; $2xy^2 \cdot -4 = -8xy^2$

$x(2x^2y - xy) = 2x^3y - x^2y$

$ab(a^2 + a + 1) = a^3b + a^2b + ab$

Polynomials, Exercise 6.

1. $7(x + 2)$

2. $3(y^2 - 5y)$

3. $x(x + 2)$

4. $2y(y - 4)$

5. $4x(x^2 + x)$

6. $6y^2(y^3 - 3y)$

7. $5x^2(x^4 + 2)$

8. $y(y^2 - y + 2)$

9. $a^2(a^3 + a + 1)$

10. $xy(x - y)$

11. $x(x^4 + 2xy + y)$

12. $x^2y(3x + 2y)$

13. $4xy^2(x^2 + 2)$

14. $7x^5(xy + y)$

15. $-pq(3p^2 - pq)$

To multiply polynomials by polynomials

Step 1: Multiply each term in the first polynomial to every term in the second polynomial.
Step 2: Collect like terms.

For example, in $(2x - 3) (4x^2 + 5x - 1)$ the terms in the first polynomial are 2x and - 3. Begin with the 2x. Multiply the 2x to the first term of the second polynomial, $4x^2$, and write the result down:

$$2x \cdot 4x^2 = 8x^3$$

$$(2x - 3) (4x^2 + 5x - 1) = 8x^3$$

Next multiply the 2x to the second term of the second polynomial, 5x, and write this result next to the $8x^3$. Then multiply the 2x to the last term of the second polynomial, -1, and write it down:

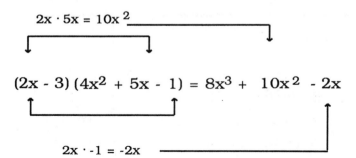

$$2x \cdot 5x = 10x^2$$

$$(2x - 3) (4x^2 + 5x - 1) = 8x^3 + 10x^2 - 2x$$

$$2x \cdot -1 = -2x$$

We are finished with 2x. Now use -3. Multiply -3 to the $4x^2$, 5x, and -1. writing down the results each time:

$$-3 \cdot 4x = -12x^2$$

$$(2x - 3) (4x^2 + 5x - 1) = 8x^3 + 10x^2 - 2x - 12x^2 - 15x + 3$$

$$-3 \cdot -1$$

$$-3 \cdot 5x = -15x$$

The preceding explains the order in which to multiply. In practice all the multiplications are done mentally and the results are written in just one line. The entire process without the arrows looks like this:

$$(2x - 3)(4x^2 + 5x - 1) = 8x^3 + 10x^2 - 2x - 12x^2 - 15x + 3$$
$$= 8x^3 - 2x^2 - 17x + 3 \quad \text{Collect like terms.}$$

Polynomials, Exercise 7. Multiply.

1. $(2x - 3)(5x^2 + x + 4)$ Hint: use the example as a guide.
2. $(3y + 1)(2y^2 + y - 4)$
3. $(x - 1)(7x^2 + 3x - 2)$
4. $(a + 1)(a^2 - 5a + 1)$
5. $(2x + 3)(5x + 1)$ Hint: first multiply 2x to each term in (5x + 1), then multiply 3 to (5x + 1). Collect like terms.

The next section offers a better way to do Problem 5.

Multiplying a Binomial by a Binomial: F.O.I.L.

As in the previous section, each term in the first parentheses must multiply each term in the second parentheses. This is best accomplished by the F.O.I.L. Method.

Each letter in F.O.I.L. represents a pair of terms to be multiplied together.

•Example 1:

F.O.I.L. $(2x - 3)(4x + 5)$.

F stands for the <u>first terms</u> of each binomial.
Multiply the first terms and write the result down.

F

Step 1: $(2x - 3)(4x + 5) = 8x^2$

O stands for the <u>outer terms</u> . Multiply the outer terms together and write the result after $8x^2$.

O

Step 2: $(2x - 3)(4x + 5) = 8x^2 + 10x$

49

I stands for <u>inner terms</u>. Multiply the inner terms.

$$\cdot\,I$$

Step 3: $(2x - 3)(4x + 5) = 8x^2 + 10x - 12x$

with the F O I labels above

L stands for the <u>last terms</u> of each binomial.

$$L$$

Step 4: $(2x - 3)(4x + 5) = 8x^2 + 10x - 12x - 15$

with the F O I L labels above

$$= 8x^2 - 2x - 15 \quad \text{Collect terms}$$

The above example breaks the FOIL Method into steps to demonstate the order in which to multiply. In practice the whole FOIL would be done in one line below the problem. The entire process looks like this:

$(2x - 3)(4x + 5)$

$= 8x^2 + 10x - 12x - 15$ By FOIL

$= 8x^2 - 2x - 15$ Collecting like terms.

•Example 2:

Multiply $(y - 8)(3y - 2)$.

$$O$$
$$F$$
$(y - 8)(3y - 2)$
$$I$$
$$L$$

$$\text{F. O. I. L.}$$
$= 3y^2 - 2y - 24y + 16$

$= 3y^2 - 26y + 16 \quad \text{Collecting terms.}$

•Example 3:

Multiply $7(x + 2)(x + 3)$.

$= 7(x^2 + 3x + 2x + 6)$ By FOIL

$= 7(x^2 + 5x + 6)$ Collecting like terms.

$= 7x^2 + 35x + 42$ Distribute the 7.

•Example 4:

> Use F.O.I.L. to multiply: a. $(2x + y) (x + 3y)$, and b. $(x + 5) (y +2)$.
>
> a. $(2x + y) (x + 3y)$ b. $(x + 5) (y +2)$
> $= 2x^2 + 6xy + xy + 3y^2$ FOIL. $=xy + 2x + 5y + 10$
> $=2x^2 + 7xy + 3y^2$ Collect terms.
>
> In Part b, there are no like terms to collect.

Polynomials, Exercise 8. Multiply using F.O.I.L.

1. $(2x - 3) (4x + 5)$ Hint: Example 1 .
2. $(3y - 4) (2y + 5)$
3. $(x + 3) (x + 2)$ Hint: $x \cdot x = x^2$
4. $(y - 2) (y + 5)$
5. $(5y + 2) (y - 4)$
6. $(x + 1) (x + 5)$
7. $(p + 6) (p - 2)$
8. $(2x - 3) (x - 2)$
9. $(y - 3) (y - 4)$
10. $(x + 1) (x + 1)$
11. $7(x + 1) (x + 1)$ Hint: First FOIL, then distribute 7 as in Example 3.

12. $(3y - 1) (y - 2)$
13. $(p + 2) (p - 2)$
14. $(2x + 1) (2x + 3)$
15. $(5y - 2) (2y - 1)$
16. $2(y + 1) (y + 2)$
17. $5(p - 2) (p - 2)$
18. $7(y + 2) (y - 3)$
19. $(x + y) (x + y)$ Example 4.
20. $(x + y) (x - y)$
21. $(x + 2y) (x + y)$
22. $(3a + b) (a + 2b)$

Difference of Squares

$(x + 6)$ and $(x - 6)$ are examples of binomials that can be multiplied together by a faster method than FOIL . This alternative method, called the **Difference of Squares**, only works in problems like $(x + 6) (x - 6)$ or $(2y - 3) (2y + 3)$ where the binomials are almost identical except one has a '+'and the other has a '- '.

To Multiply by the Difference of Squares Method

Square the first expression and subtract the square of the second expression.

For example, in $(x + 6) (x - 6)$ the first expression in both binomials is x. The second expression in both binomials is 6. Then:

$$(x + 6) (x - 6) = x^2 - 6^2 = x^2 - 36 .$$

•More examples:

1. $(x - 9)(x + 9) = x^2 - 9^2 = x^2 - 81$
2. $(2y - 3)(2y + 3) = (2y)^2 - 3^2 = 4y^2 - 9$
3. $(5x + 4y)(5x - 4y) = (5x)^2 - (4y)^2 = 25x^2 - 16y^2$

In general, $(A + B)(A - B) = A^2 - B^2$

The procedure is called the Difference of Squares because the result is always a subtraction (or difference) of two squares.

If you wish to see why this method works, consider Example 1 by the FOIL Method:

F.   O.   I.   L.

$(x + 9)(x - 9)$ $= x^2 - 9x + 9x - 81$.

$= x^2 - 81$ Since $-9x + 9x = 0x = 0$.

The **O** and **I** parts cancel each other out.
The **F** supplies the first expression squared, namely x^2.
The **L** supplies the negative sign and the last expression squared, -81.

Polynomials, Exercise 9. Multiply each of the following by (a) the FOIL Method and (b) the Difference of Squares Method.

1. $(x + 9)(x - 9)$
2. $(y - 10)(y + 10)$

3. $(2x + 1)(2x - 1)$
4. $(a + b)(a - b)$

Multiply using the Difference of Squares Method only.

5. $(x + 1)(x - 1)$
6. $(p + 7)(p - 7)$
7. $(p - 7)(p + 7)$
8. $(3x + 4)(3x - 4)$

9. $(4y - 5)(4y + 5)$
10. $(x + 4y)(x - 4y)$
11. $(x^2 + 11)(x^2 - 11)$
12. $(A + B)(A - B)$

The Square of a Binomial

$(x - 5)^2$, $(y + 10)^2$, and $(3x - 5)^2$ are examples of Squares of Binomials. The power of 2 indicates that the binomial is multiplied to itself. For instance, $(x - 5)^2 = (x - 5)(x - 5)$, two factors of $(x - 5)$.

A binomial can be multiplied to itself either by the FOIL Method or by an alternate method called the Square of a Binomial Method.

To understand how the Square of a Binomial Method works, first consider $(x - 5)^2$ by the FOIL Method:

$$\begin{array}{c} \text{F}\quad\ \text{O}\quad\ \ \text{I}\qquad \text{L} \\ (x - 5)^2 = (x - 5)(x - 5) = x^2 - 5x - 5x + 25 \\ = x^2 - 10x + 25 \end{array}$$

The O and I parts of FOIL are identical so that -5x appears twice. In other words, the -5x *is doubled* to make the middle term -10x. This observation leads to the alternative method.

Square of Binomial Method

Step 1: Square the first term and write the result down.
Step 2: Multiply the first term to the last term and double the result. In other words, 2· (first term) (last term).
Step 3: Square the last term.

•Example:

$$\begin{array}{l} (x - 5)^2 = x^2 + 2(x)(-5) + (-5)^2 \\ = x^2 - 10x + 25 \qquad \text{Same answer as above by FOIL.} \end{array}$$

•More examples:

$$(y + 10)^2 = y^2 + 2(y)(10) + 10^2 = y^2 + 20y + 100$$
$$(p - 4)^2 = p^2 + 2(p)(-4) + (-4)^2 = p^2 - 8p + 16$$
$$(7x + 2y)^2 = (7x)^2 + 2(7x)(2y) + (2y)^2 = 49x^2 + 28xy + 4y^2$$

In general, $(A + B)^2 = A^2 + 2AB + B^2$

Polynomials, Exercise 10. Multiply by (a.) the F.O.I.L. Method and by (b.) the Square of Binomial Method.

1. $(x - 3)^2$
2. $(y + 9)^2$
3. $(2x + 1)^2$
4. $(A + B)^2$

Multiply by the Square of Binomial Method only.

5. $(x - 7)^2$
6. $(x + 7)^2$
7. $(4y + 5)^2$
8. $(x + 5y)^2$
9. $(x^3 + 11)^2$
10. $(A - B)^2$

Factoring Polynomials

Factors and Products

The answer to a multiplication problem is called the product. The expressions being multiplied together are called factors. For example, in $2 \cdot 7 = 14$, the product is 14 and the factors are 2 and 7.

•More examples:

> In $2 \cdot 3 = 6$, the product is 6. The factors are 2 and 3.
>
> In $x^2 \cdot x^3 = x^5$, the product is x^5. The factors are x^2 and x^3.
>
> In $2(y + 7) = 2y + 14$, the product is $2y + 14$.
> The factors are 2 and $y + 7$.
>
> In $(x + 5)(x + 2) = x^2 + 2x + 5x + 10$ (by F.O.I.L.), the product is $x^2 + 2x + 5x + 10$. The factors are $(x + 5)$ and $(x + 2)$.
>
> In $(y + 3)(y - 2) = y^2 + y - 6$, the product is $y^2 + y - 6$. The factors are $(y + 3)$ and $(y - 2)$.

The **goal** of this chapter is to teach you how to find the factors when only the products are known. If the product is a polynomial, the process of finding its factors is called **factoring the polynomial**. To factor polynomials, we need a concept known as *the greatest common factor* .

The Greatest Common Factor (GCF)

The **greatest common factor** of two or more numbers is the largest number that divides each number evenly.

To find the GCF of two or more numbers:

Step 1: Write each number as a product of primes.
Step 2: Identify the factors common to each number.
Step 3: Multiply the common factors together.

•Example:

Find the GCF of 24, 36, and 60.

Step 1:

24	36	60
$=4\cdot6$	$=6\cdot6$	$=6\cdot10$
$=2\cdot2\cdot2\cdot3$	$=2\cdot3\cdot2\cdot3$	$=2\cdot3\cdot2\cdot5$

Step 2: Two factors of 2 and one factor of 3 are common to each number:

$$24 = 2\cdot\mathbf{2\cdot2\cdot3} \qquad 36 = \mathbf{2\cdot3\cdot2}\cdot3 \qquad 60 = \mathbf{2\cdot3\cdot2}\cdot5$$

Hence GCF = $2\cdot2\cdot3$ = 12.
12 is the largest number that divides evenly into 24, 36, & 60.

The GCF of Two or More Variable Expressions

The greatest common factor of two or more variable expressions is the largest collection of variables common to each expression.

For example, the expressions x^3y, x^2y^2 and x^2yz have two factors of x and one factor of y in common because:

$$x^3y = x\,\mathbf{x}\,\mathbf{x}\,\mathbf{y} \qquad x^2y^2 = \mathbf{x}\,\mathbf{x}\,\mathbf{y}\,y \qquad x^2yz = \mathbf{x}\,\mathbf{x}\,\mathbf{y}\,z$$

Hence, the GCF = x x y = x^2y. A faster method is described next.

To find the GCF of Two or More Variable Expressions

Use each variable the fewest number of times it appears in any one expression.

•Example 1:

Find the GCF of x^3y, x^2y^2 and x^2yz.

x appears fewest in either x^2y^2 or x^2yz (2 factors of x).
y appears fewest in either x^3y or x^2yz (1 factor of y)
z appears fewest in either x^3y or x^2y^2 (0 factors of z).

Hence the GCF = x^2y.

•Example 2:

> Find the GCF of $x^2y(x + 3)$ and $xy(x + 3)^2$.
>
> | | |
> |---|---|
> | x appears fewest in $xy(x + 3)^2$ | (1 factor of x). |
> | y appears fewest in either expression | (1 factor of y). |
> | (x + 3) appears fewest in $x^2y(x + 3)$ | [1 factor of (x + 3)]. |
>
> Hence, the GCF = $xy(x + 3)$.

•Example 3:

> Find the GCF of $30a^2b$, $18ab^3$, and $21a^3b^2$.
>
> | $30a^2b$ | $18ab^3$ | $21a^3b^2$ |
> |---|---|---|
> | $= 3 \cdot 10a^2b$ | $= 2 \cdot 9ab^3$ | $= 3 \cdot 7a^3b^2$ |
> | $= 3 \cdot 2 \cdot 5a^2b$ | $= 2 \cdot 3 \cdot 3ab^3$ | $= 3 \cdot 7a^3b^2$ |
>
> GCF = $3ab$

•Example 4:

> Find the GCF of:
> a. 15 and 8
> b. $4x$ and $13y^3$
>
> a. The GCF of 15 and 8 is 1, since 1 is the only number that divides both 15 and 8.
>
> b. The GCF of $4x$ and $13y^3$ is 1, because they have no factors in common other than 1.

Factoring, Exercise 1. In the following, identify the product and the factors.

1. $7 \cdot 11 = 77$

2. $x^5 \cdot x^3 = x^8$

3. $4(y - 3) = 4y - 12$

4. $(x + 6)(x - 2) = x^2 - 2x + 6x - 12$

5. $42 = 7 \cdot 6$

6. $3x + 6 = 3(x + 2)$

7. $y^2 - 25 = (y + 5)(y - 5)$

8. $(2x + 1)(x + 1) = 2x^2 + 3x + 1$

9. What is the goal of this chapter?

Factoring, Exercise 2. Find the GCF.

1. 18, 24
2. 42, 70, 28
3. 4, 9
4. x^2, x^3, x^7
5. y^{13}, y^4, y^9
6. x^3, y
7. xy, y^2x, x^2y
8. p^2q^3, pq^7, p^7q^2
9. xy^2, 5x, 4

10. 4x, $8x^2$, $12x^3$
11. $9y^2$, $6y^5$, $12y^3$
12. $25x^3y^2$, $15x^2y^5$, $20x^7y$
13. 3x, 12, 15, 6
14. x^2, 4x, 3
15. $4x^2$, 4, $8x^2$
16. 4(x - 2), 2x(x - 2)
17. y^2(2y - 1), 3(2y - 1)
18. $xy(x + 3)^2$, $x^2(x + 3)$

Missing Factors and the Distributive Law

The greatest common factor and the Distributive Law will be used together in the next section to discover the factors of polynomials. The *missing factors* concept, explained below, will be used as you proceed through the next section.

For example in 8 = (2) ( ), the product is 8 and the factor is 2. The other factor, which is missing, must be 4, since 8 = (2) (4).

• More examples:

Determine the missing factors in the following:

a. 7 = (-1) ( )
b. 6x = ( ) (3x)
c. (11) ( ) = $11y^2$

d. x^5 = (x^3) ( )
e. a^2 = ( ) (a)
f. $15x^2y^2$ = ($3xy^2$) ( )

a. The missing factor is -7: since, 7 = (-1) (-7)
b. The missing factor is 2; since, 6x = 2 · 3x
c. y^2 is missing; since, 11 · y^2 = $11y^2$.
d. x^2 is missing; since, x^5 = (x^3) (x^2) by the Product Rule.
e. a^2 = (a) (a)
f. $15x^2y^2$ = ($3xy^2$) (5x)

The Distributive Law with Missing Factors

When factors are missing inside parentheses, the Distributive Law can be used to find them.

•Example 1:

> Find the missing numbers: $2(+) = 2 \cdot 7 + 2 \cdot 3$
>
> Solution: The right side of the equal sign shows the 2 distributed to both 7 and 3. Therefore, the missing numbers are 7 and 3.
>
> $2(7+3) = 2 \cdot 7 + 2 \cdot 3$

•Example 2:

> Use the Distributive Law to fill in the missing expressions:
>
> $2(+) = 2x + 10$
>
> Solution: Ask yourself, these questions:
> 1. 2 multiplied by what is 2x?
>   2 multiplied by x is 2x. Therefore, x is the first missing expression.
> 2. 2 multiplied by what is 10?
>   2 multiplied by 5 is 10. Therefore, 5 is the second missing expression.
>
> Hence, the missing expressions are x and 5.
>
> $2(x + 5) = 2x + 10.$

•Example 3:

> Fill in the missing expressions:
>
> $7x - 14$
> $= 7()$
>
> Solution:
> 1. 7 multiplied by what is 7x?
>   Answer is x, since $7 \cdot x = 7x$
> 2. 7 multiplied by what is -14?
>   Answer -2, since $7 \cdot -2 = -14$
>
> Therefore, the missing expressions are x and -2:
>
> $7x - 14$
> $= 7(x - 2)$

Factoring, Exercise 3. Fill in the missing factors.

1. $20 = (4)(\quad)$
2. $-10 = (\quad)(2)$
3. $14x = (7)(\quad)$
4. $x^2 = (x)(\quad)$

5. $y^2 = (\quad)(y)$
6. $x^3 = x^2(\quad)$
7. $y^5 = y(\quad)$
8. $8x^2 = (\quad)(4x)$

9. $6xy^2 = (3x)(\quad)$
10. $25a^3 b^2 = (5a^2 b^2)(\quad)$
11. $16p^2 q^2 = (\quad)(8p)$
12. $5x^6 = (-5x^2)(\quad)$

Fill in the missing expressions:

13. $5(\quad+\quad) = 5 \cdot 8 + 5 \cdot 9$

14. $3 \cdot 11 + 3 \cdot 2 = 3(\quad+\quad)$

15. $4x + 4 \cdot 8$
$= 4(\quad+\quad)$

Answer the questions, then fill in the missing expressions.

16. 7 multiplied by what is $7x$?

   7 multiplied by what is 21?

   $7(\quad+\quad) = 7x + 21$.

17. -3 multiplied by what is $-3y$?

   -3 multiplied by what is 6?

   Hence, $-3y + 6$

   $\qquad = -3(\qquad)$

18. x multiplied by what is x^2?

   x multiplied by what is $10x$?

   Hence $x^2 + 10x$

   $\qquad = x(\qquad)$

Fill in the missing factors by asking yourself appropriate questions.

19. $8x + 16$
   $= 8(\qquad)$

20. $-4y - 20$
   $= -4(\qquad)$

21. $xy + 2x$
   $= x(\qquad)$

Factoring Out the Greatest Common Factor

If the terms of a polynomial have a greatest common factor (GCF) larger than 1, then the GCF can be factored out.

For example, $5x^3 - 30x^2 + 25x - 15$ is a polynomial with a GCF of 5, since 5 divides evenly into 5, -30, 25 and -15.

To Factor Out the GCF (Method 1)

Step 1: Determine the GCF and write it down followed by parentheses.
Step 2: Fill in the missing expression of the parentheses.
Step 3: Check the result by distributing the GCF back through the parentheses.

•Example 1:

Factor out the GCF: $5x^3 - 30x^2 + 25x - 15$

Step 1: $5($ $)$

Step 2:   5 multiplied by what is $5x^3$?   Answer, x^3.

            5 multiplied by what is $-30x^2$?   Answer, $-6x^2$.

            5 multiplied by what is $25x$?   Answer, $5x$.

            5 multiplied by what is -15?   Answer, -3.

The missing factors are x^3, $-6x^2$, $5x$, and -3.

Therefore, $5(x^3 - 6x^2 + 5x - 3)$ is the desired result.

Step 3: Check the solution by distributing the 5:

$5(x^3 - 6x^2 + 5x - 3) = 5x^3 - 30x^2 + 25x - 15$ which is the original polynomial.

The questions in Step 2 and the check in Step 3 are usually done mentally and are not written down. The entire process should look like:

$$5x^3 - 30x^2 + 25x - 15$$
$$= 5(x^3 - 6x^2 + 5x - 3)$$

•Example 2:

Factor $xy^3 - 7x^2y^2$.

The GCF of the terms is xy^2, because x appears fewest in xy^3 and y appears fewest in $7x^2y^2$.

Hence, $xy^3 - 7x^2y^2$

Step 1: $= xy^2($ $)$

Step 2: xy^2 multiplied by what is xy^3?

            Answer, y; since $xy^2 \cdot y = xy^3$.

            xy^2 multiplied by what is $-7x^2y^2$?

            Answer $-7x$; since $xy^2 \cdot -7x = -7x^2y^2$.

            Therefore, $xy^2(y - 7x)$ is the desired result.

Step 3:   By the Distributive Law, $xy^2(y - 7x) = xy^3 - 7x^2y^2$.

When the GCF is showing itself in each term.

If each individual term is written as the GCF multiplied by another expression, the GCF can be easily factored out.

For example, in 6x + 6y - 6p, the GCF of the terms is 6. Moreover, each individual term is written as 6 multiplied by another expression.

To Factor Out the GCF (Method 2)

Step 1: Determine the GCF and write it down followed by parentheses.
Step 2: Fill in the parentheses by recopying each term without its GCF component.

•Example 1:

$$6x + 6y - 6p$$

$$= 6(x + y - p) \qquad \text{Combining Steps 1 and 2.}$$

The result checks by the Distributive Law.

•Example 2:

Factor $x(x + 2) + y(x + 2) - 5(x + 2)$.

Solution: The GCF of the terms is $(x + 2)$.

$$x(x + 2) + y(x + 2) - 5(x + 2)$$

Step 1: $= (x + 2) (\qquad)$

Step 2: $= (x + 2) (x + y - 5)$

•Example 3:

Factor $5x + 2y + 13$.

There are no common factors. The GCF = 1. The polynomial cannot be factored by either method.

Factoring, Exercise 4. Factor.

Use Method 1

1. $4x + 12$
2. $2x - 10$
3. $5x + 2$
4. $6x - 6$
5. $3x^2 + 6x$
6. $5y^2 - 10y^3$
7. $15x^2 + 2y^4$

Use Method 2

11. $5x + 5y + 5z$
12. $3x^2 - 3xy + 3y$
13. $ax + ay + az$
14. $4x + 7y + 4z$
15. $x(x - 1) + y(x - 1) + 3(x - 1)$
16. $xy(y + 1) + 4x(y + 1) - 5(y + 1)$
17. $x(x - 3) + 2(x - 3)$

8. $7p^6 - 3p^3$

9. $4x^4 + 2x$

10. $2x^2y^2 + 4x^2 - 6x + 10$

18. $x(x + 7) + 2(x - 5)$

19. $x(x + 2) + 2(x + 2)$

20. $x(x - 3) + x - 3$

Factoring Polynomials Which Have Four Terms

The goal in this section is to help you find the factors when the product is a four term polynomial. For example in:

$$(x + 5)(y + 2) = xy + 2x + 5y + 10 \text{ (by FOIL)},$$

the product is $xy + 2x + 5y + 10$ and the factors are $(x + 5)$ and $(y + 2)$.

The factors of a four term polynomial can often be discovered through a process called **grouping**. Not every four term polynomial can be factored.

To Factor by Grouping

Step 1: Factor out the greatest common factor of all four terms.

Step 2: Take a common factor out of the first two terms. Also take a common factor out of the last two terms.

Step 3: Factor out the common parentheses as in the previous section (see Method 2, Example 2).

Step 4: Check. FOIL the factors and compare the result to the original problem.

•Example 1:

Find the factors of $xy + 2x + 5y + 10$.

Step 1: The four terms have no common factor other than 1. Skip Step 1.

Step 2: x is a common factor of the first two terms.
5 is a common factor of the last two terms.

Hence,
$$xy + 2x + 5y + 10$$ Step 2. Because $xy + 2x = x(y + 2)$
$$=x(y + 2) + 5(y + 2)$$ and $5y + 10 = 5(y + 2)$.

$$= (y + 2)(x + 5)$$ Step 3. Because $9y + 2)$ is the GCF.

The factors are $(y + 2)(x + 5)$.   It does not matter in what order they appear. Check:

$$(y + 2)(x + 5) = xy + 5y + 2x + 10$$
$$= xy + 2x + 5y + 10$$

•Example 2:

Factor $2y^2 + y - 6y - 3$.

    Step 1: Skip, because the GCF of all four terms is 1.

$2y^2 + y - 6y - 3$
$= y(2y + 1) - 3(2y + 1)$ Step 2. Since $-3(2y + 1) = -6y - 3$.
$= (2y + 1)(y - 3)$ Step 3.

Check the result : $(2y + 1)(y - 3) = 2y^2 - 6y + y - 3$
$= 2y^2 + y - 6y - 3$

•Example 3:

Factor $3x^3 - 12x^2 + 6x^2 - 24x$.

$3x^3 - 12x^2 + 6x^2 - 24x$.
$= 3x[x^2 - 4x + 2x - 8]$ Step 1. GCF $= 3x$.
$= 3x[x(x - 4) + 2(x - 4)]$ Step 2. $x^2 - 4x = x(x - 4)$ and
        $2x - 8 = 2(x - 4)$.
$= 3x[x(- 4)(x + 2)]$ Step 3. GCF $= (x - 4)$.
$= 3x(x - 4)(x + 2)$

Factoring, Exercise 5.

Factor by grouping.

1. $xy + 2x + 3y + 6$ 6. $4y^2 + 20y + 20y + 100$

2.. $x^2 - 5x + 4x - 20$ 7. $x^2 + 2x + 3x + 9$ Hint: Read the answer.

3. $xy + 7x - 2y - 14$ Hint: factor -2 out of the last two terms.

4. $3y^2 - 4y - 9y + 12$ Hint: factor a -3 out of the last two terms.

5. $5xy + 10x + 15y + 30$ Hint: Example 3.

Factoring Trinomials

The goal in this section is to help you find the factors when the product is a trinomial. For example, in:

$$(2x + 1)(x + 3) = 2x^2 + 7x + 3 \text{ (by FOIL)},$$

the product is $2x^2 + 7x + 3$ and the factors are $(2x + 1)$ and $(x + 3)$. In other words, the trinomial product has binomial factors.

The binomial factors of a trinomial can often be discovered through a process called **trial and error**. Not every trinomial can be factored, however.

To Factor Trinomials by Trial and Error

Build two binomials in such a way that, when they are multiplied together by FOIL, the result is the original trinomial. This process can be outlined in steps:

> **Step 1:** Draw parentheses
> **Step 2:** Fill in the missing factors to get the first term.
> **Step 3:** Fill in the missing factors to get the last term.
> **Step 4:** Check by FOIL.

•Example 1:

Find the binomial factors of $2x^2 + 7x + 3$.

Note: The <u>first term</u> in the trinomial is $2x^2$. The <u>last term</u> in the trinomial is 3. The <u>middle term</u> in the trinomial is $7x$.

Step 1: Draw two pairs of parentheses big enough to hold two binomials, then underline the first position in each parentheses:

$$2x^2 \; + \; 7x \; + \; 3$$
$$(\underline{} \qquad)(\underline{} \qquad)$$

Step 2: These underlined positions multiplied together will provide the <u>first term</u> in the trinomial, namely $2x^2$. To fill in these underlined positions, ask yourself this question: What multiplied by what equals $2x^2$?

Answer: $2x$ multiplied by x equals $2x^2$. Therefore the underlined positions are $2x$ and x.

$$2x^2 \; + \; 7x \; + \; 3$$
$$(\underline{2x} \qquad)(\underline{x} \qquad)$$

Step 3: The second positions of each parentheses, when multiplied together, must provide the <u>last term</u> of the trinomial, namely 3. Ask: What multiplied by what equals 3? Answer: 3 times 1 equals 3.

$$2x^2 + 7x + 3$$
$$(2x + 3)(x + 1)$$

Step 4: Multiply these binomials together by FOIL to see if the result is the original trinomial.

$$
\begin{array}{ll}
& \quad\ \ \text{F} \quad\ \text{O} \quad \text{I} \quad\ \text{L} \\
(2x + 3)(x + 1) & = 2x^2 + 2x + 3x + 3 \\
& = 2x^2 + 5x + 3
\end{array}
$$

Alas, the original trinomial was $2x^2 + 7x + 3$, not $2x^2 + 5x + 3$. Therefore, the placement of 3 and 1 is not correct.

Try **switching** the 3 and the 1. That is, put the 1 in the first parentheses and 3 in the second:

$$
\begin{array}{ll}
(2x + 1)(x + 3) & = 2x^2 + 6x + x + 3 \\
& = 2x^2 + 7x + 3 \qquad \text{Eureka!}
\end{array}
$$

Same as the original trinomial. The factors of $2x^2 + 7x + 3$ are:

$$(2x + 1)(x + 3)$$

In practice, all these words in Example 1 are not written down. The entire process looks like this:

First Try	Second Try
$2x^2 + 7x + 3$	$2x^2 + 7x + 3$
$(2x + 3)(x + 1)$	$(2x + 1)(x + 3)$
Check by FOIL	Check by FOIL
$2x^2 + 2x + 3x + 3$	$2x^2 + 6x + x + 3$
$2x^2 + 5x + 3$	$2x^2 + 7x + 3$
Alas!	Eureka!

Therefore, $2x^2 + 7x + 3 = (2x + 1)(x + 3)$.

•Example 2:

Factor $3y^2 - 2y - 5$.

$$3y^2 - 2y - 5$$
$$(3y \qquad)(y \qquad) \qquad \text{Steps 1 and 2, since } 3y \cdot y = 3y^2.$$

Step 3: The last term in the trinomial is -5. Therefore, the missing factors are either +5 and -1 , or they are -5 and +1, because both pairs of numbers multiply to -5.

When confronted with a choice of missing factors, simply pick one pair or the other and try it.

Let's try +5 and -1, with +5 in the first parentheses and -1 in the second:

$$3y^2 - 2y - 5$$
$$(3y + 5)(y - 1) \qquad = 3y^2 - 3y + 5y - 5 \qquad \text{By FOIL.}$$
$$= 3y^2 + 2y - 5$$

Alas, the original trinomial was $3y^2 - 2y - 5$, not $3y^2 + 2y - 5$. The only part that is not the same is the sign of the middle term: it should be -2y and not +2y. The situation is easily remedied by switching the signs inside the binomials.

$$3y^2 - 2y - 5$$
$$(3y - 5)(y + 1) \qquad \text{Switch the signs.}$$
$$3y + 3y - 5y - 5$$
$$3y - 2y - 5. \qquad \text{Eureka!}$$

Therefore, the factors of $3y^2 - 2y - 5$ are:

$$(3y - 5)(y + 1)$$

Example 2 without all the words looks like:

Factor $3y^2 - 2y - 5$.

First Try	Second Try
$(3y + 5)(y - 1)$	$(3y - 5)(y + 1)$
Check	Check
$3y^2 - 3y + 5y - 5$	$3y^2 + 3y - 5y - 5$
$3y^2 + 2y - 5$	$3y^2 - 2y - 5$
Alas.	Eureka!

•Example 3:

Factor $10x^2 + 18xy - 4y^2$.

2 is a common factor. Take out the 2.
$$10x^2 + 18xy - 4y^2$$
$$=2(5x^2 + 9xy - 2y^2)$$

Now factor the remaining trinomial: $5x^2 + 9xy - 2y^2$.

$2(5x^2 + 9xy - 2y^2)$

$= 2(5x\quad)(x\quad)$ Step 1 and 2.

$= 2(5x - y)(x + 2y)$ Step 3.

Check by FOIL.

$2(5x^2 + 10xy - xy - 2y^2)$

$= 2(5x^2 + 9xy - 2y^2)$ It checks.

Therefore, the factors of $10x^2 + 18xy - 4y^2$ are:

$2(5x - y)(x + 2y)$

Example 3 without words.

$10x^2 + 18xy - 4y^2$		Check	$2(5x - y)(x + 2y)$
$2(5x^2 + 9xy - 2y^2)$			$2(5x^2 + 10xy - xy - 2y^2)$
$2(5x - y)(x + 2y)$			$2(5x^2 + 9xy - 2y^2)$
			$10x^2 + 18xy - 4y^2$

Examples 1, 2, and 3 suggest the following hints:
1. If there is a common factor, take it out first.
2. When checking by FOIL:
    a. If the coefficient of the middle term is different than the coefficient of the original polynomial, try switching the last factors.
    b. If the middle term is only off by a sign, try switching the signs in each binomial.

Factoring, Exercise 6. Factor, then check by F.O.I.L.

1. $x^2 + 5x + 6$

2. $y^2 - 7y + 12$

3. $2x^2 + 5x + 2$

4. $5y^2 + 9y - 2$

5. $x^2 + 5x + 1$ (see below)

6. $4x^2 - 4x + 1$ (see below)

7. $4x^2 + 16x + 15$ (see below)

8. $2x^2 - 14x + 24$ (see Example 3)

9. $18x^2 + 15x + 3$ (common 3)

10. $3p^3 + 9p^2 + 6p$ (common 3p)

11. $x^2 + 14xy + 49y^2$

12. $x^2 + 2xy - 35y^2$

13. $2x^2 + 3xy + y^2$

14. $3p^2 + 4pq + q^2$

15. $3x^2 + xy + y^2$ (see answer)

16. $7a^2 - 15ab + 2b^2$

17. $5x^2 + 10xy + 5y^2$ (common 5)

18. $7x^2 - 21xy + 14y^2$ (common 7)

19. $8x^2 + 10xy + 2y^2$ (common 2)(see below)

20. $x^3 + 6x^2y + 9xy^2$ (common x)

Further hints: In problems 6 and 7, the first positions are either $(4x\quad)(x\quad)$ or $(2x\quad)(2x\quad)$. Problems 6, 7, 9, and 19 are the most challenging. Problem 5 is not factorable by this method.

Factoring Binomials

The goal in this section is to learn how to find the binomial factors when the product is a **difference-of-squares** binomial. For example, in:

$$(x + 5)(x - 5)$$
$$= x^2 - 25$$

the factors are binomials and the product, $x^2 - 25$, is an example of a difference of squares binomial.

A binomial is called a difference-of-squares binomial if: (1) There is a minus sign between the expressions; and, (2) Both expressions are squares.

For instance, $x^2 - 25$ is a difference of squares, because x^2 is a square and 25 is also a square ($25 = 5^2$). Other examples are $y^2 - 36$, $p2 - 4$, and $4x^{10} - 81y^2$.

To Find the Factors of a Difference of Squares Binomial

Rewrite each expression as a base raised to the power of two. Add the bases to get the first factor. Subtract the bases to get the second factor.

•Example 1:

Factor $x^2 - 25$.

$$x^2 - 25 \quad = x^2 - 5^2$$
$$= (x + 5)(x - 5)$$

•Example 2:

Factor $4x^{10} - 81y^2$.

$$4x^{10} - 81y^2$$
$$= (2x^5)^2 - (9y)^2 \qquad \text{By the Power Rule}$$
$$= (2x^5 + 9y)(2x^5 - 9y)$$

•Example 3:

Factor $4x^2 y - 400y$.

The GCF is 4y.
$$4x^2 y - 400y$$
$$4y(x^2 - 100)$$
$$4y(x^2 - 10^2)$$
$$4y(x + 10)(x - 10)$$

Factoring, Exercise 7. Factor.

1. $x^2 - 16$
2. $y^2 - 64$
3. $x^2 - 49$
4. $3y^2 - 27$ Common 3
5. $xy^2 - 4x$ Common x
6. $7x^2y - 700y$ See Example 3
7. $x^3 - xy^2$
8. $x^2 + 1$ Does not factor.

9. $x^2 - 1$ Hint: $1 = 1^2$
10. $7x^2 - 7$ Factor out the 7.
11. $64x^2 - 25$
12. $9x^2 - 25$
13. $45x^6 - 20y^4$ Common 5
14. $25x^4y^2 - 9p^2q^2$
15. $3x^2 - 4$

Summary of Factoring Polynomials

This chapter has shown how to factor polynomials having two, three, or four terms. The factors of these types of polynomials will be used later to solve *second degree equations* .

The outline below provides a review of which method applies to binomials, to trinomials, and polynomials having four terms. Remember, however, not all polynomials can be factored.

Step 1: If all terms have a greatest common factor other than one, then factor it out.

Step 2: If the polynomial has:
1. four terms, then try the *factoring by grouping method* .
2. three terms, then try the *trial and error method* .
3. two terms, then it will factor only if it is a *difference of squares* binomial.

For examples, refer back to the previous sections.

Factoring, Exercise 8. Factor each.

1. $x^2 + 2x + 1$
2. $x^2 - 9$
3. $x^2 + 2x + 3x + 6$
4. $3y^2 - 48$ GCF = 3
5. $2x^2 + 3x + 1$
6. $3y^2 + 6y + 3$ Step 1 first.
7. $5x^2 - 20$
8. $4y^2 + 6y + 12$
9. $3y^2 + 5y + 2$

10. $x^2 + 10$
11. $y^3 - 9y$ GCF = y
12. $x^2 + 4x + 1$
13. $x^2y - 4xy + 2y$
14. $x^3 - 4x^2 + 3x$
15. $2y^2 - 6$
16. $5x^2 + 35x + 10x + 70$
17. $7x^2 - 49x + 14$
18. $ab^2 - a$

Second Degree Equations

$y + 12 = y^2$ and $x^2 - 3x + 2 = 0$ are called **second degree equations** since the highest power on each variable is 2.

Solving an equation means finding the values that make the equation true. To solve second degree equations, we need a concept called the **Zero Product Rule**.

Zero Product Rule: if the product of a multiplication is zero, then either the first factor equals zero or the second factor equals zero (or both factors are zero).

•Example 1:

> Fill in the missing factor: (8) (   ) = 0.
>
> > Solution: The product is zero. Therefore, by the Zero Product Rule, the second factor must be zero.
> >
> > $$(8)(0) = 0$$

When the factors are polynomials, the Zero Product Rule allows us to find the solution to second degree equations.

To Solve Second Degree Equations

Step 1: Get zero on one side of the equal sign.
Step 2: Factor the polynomial.
Step 3: Apply the Zero Product Rule.

•Example 1:

> Solve $3x^2 + 2x - 1 = 0$.
>
> > Solution: Zero is already on one side of the equal sign.
> >
> > $$3x^2 + 2x - 1 = 0$$
> >
> > Step 2: $(3x - 1)(x + 1) = 0$ Factoring the trinomial.
> >
> > Step 3: The product is zero and the factors are $(3x - 1)$ and $(x + 1)$. The Zero Product Rule says that either the first factor equals zero or the second factor equals zero. In other words:
> >
> > $$\text{either} \quad 3x - 1 = 0 \quad \text{or} \quad x + 1 = 0$$

These two smaller equations can be solved using methods introduced in Chapter 3.

$$3x - 1 = 0 \qquad \text{or} \qquad x + 1 = 0$$
$$3x = 1 \qquad \text{or} \qquad x = -1 \qquad \text{Addition principle.}$$
$$\frac{3x}{3} = \frac{1}{3} \qquad \text{or} \qquad x = -1$$
$$x = \frac{1}{3} \qquad \text{or} \qquad x = -1$$

The Zero Product Rule allows the original equation to be broken up into small pieces. The entire process without all the extra words looks like this:

$$3x^2 + 2x - 1 = 0$$
$$(3x - 1)(x + 1) = 0$$

either $\quad 3x - 1 = 0 \qquad$ or $\qquad x + 1 = 0$

$$3x = 1 \qquad \text{or} \qquad x = -1$$
$$x = \frac{1}{3} \qquad \text{or} \qquad x = -1$$

Example 1 shows that $x = \frac{1}{3}$ and $x = -1$ are values that make $3x^2 + 2x - 1 = 0$ true. We can check that $x = -1$ is a solution by replacing both xs with -1:

$$3x^2 + 2x - 1 = 0$$
$$3(-1)^2 + 2(-1) - 1 = 0$$
$$3 \cdot 1 - 2 - 1 = 0$$
$$0 = 0 \qquad \text{True.}$$

A similar (but more difficult) check would show that $x = \frac{1}{3}$ is also a solution.

•Example 2:

Solve $x^2 = -7x$.

Get zero on the right side of the equation by adding $7x$ to both sides.

$$x^2 + 7x = -7x + 7x$$
$$x^2 + 7x = 0 \qquad \text{Since } -7x + 7x = 0x = 0.$$
$$x(x + 7) = 0 \qquad \text{Factor out the common } x.$$

either $x = 0 \quad$ or $\; x + 7 = 0 \quad$ Zero Product Rule.

$$x = 0 \quad \text{or} \quad x = -7$$

•Example 3:

Solve $2 = x + 3x^2$.

Get zero on the left side by subtracting 2 from both sides.

$$2 = x + 3x^2$$
$$2 - 2 = x + 3x^2 - 2$$
$$0 = x + 3x^2 - 2$$

To factor this trinomial, it must first be rewritten in descending order. Move the $3x^2$ term to the front.

$$0 = x + 3x^2 - 2$$
$$0 = 3x^2 + x - 2$$
$$0 = (3x - 2)(x + 1) \qquad \text{Factors.}$$

either $3x - 2 = 0$ or $x + 1 = 0$ Zero Product Rule.

$$3x = 2 \quad \text{or} \quad x = -1$$
$$x = \frac{2}{3} \quad \text{or} \quad x = -1$$

•Example 4:

Solve $-x^2 + 5x - 6 = 0$.

Zero is already on one side. However, trinomials are too difficult to factor when the first term is negative. Multiply both sides by -1 to get:

$$-x^2 + 5x - 6 = 0$$
$$-1[-x^2 + 5x - 6] = 0 \, (-1)$$
$$x^2 - 5x + 6 = 0$$
$$(x - 2)(x - 3) = 0$$

either $x - 2 = 0$ or $x - 3 = 0$

$$x = 2 \quad \text{or} \quad x = 3$$

Factoring, Exercise 9. solve.

1. $2x^2 + 3x + 1 = 0$

2. $x^2 - 7x + 12 = 0$

3. $4x^2 + x = 0$

4. $x^2 - 9 = 0$

5. $2y^2 + 5y = -3$

6. $x^2 = -3x$

7. $x^2 = 5x - 6$

8. $6x + x^2 + 9 = 0$ Hint: Descending order.

9. $-x^2 - 3x + 10 = 0$

10. $0 = x^2 - 5x + 4$

11. $0 = 4y^2 + 8y$

12. $0 = 2x^2 - 50$

Review

Chapter 5: Use the rules of exponents to simplify the following.

1. $x^8 \cdot x^3$

2. $(z^6)^3$

3. $\dfrac{y^{10}}{y}$

4. p^0

5. x^{-5}

6. $(xy)^4$

7. $y^4 \cdot y$

8. $\dfrac{z^6}{z^5}$

9. $(2^4)^5$

10. $\dfrac{4^5}{4^4}$

11. 19^0

12. $x^{-7} \cdot x^{-2}$

13. $\dfrac{y^5}{y^{10}}$

14. $(x^{-2})^5$

15. $y^{-5} \cdot y$

16. $\dfrac{z}{z^{-5}}$

Chapter 6: State whether the following are monomials, binomials, or trinomials. Also state the degree of each polynomial.

1. $x^5 - 4$

2. $y - x + 2$

3. $14xy$

4. $x^2 y^4 + z$

Collect like terms.

5. $x^2 + 6x + 2x - 1$

6. $4x + 3 - 9x - 9$

7. $yx^2 + z - 4yx^2$

8. $5y^4 + 7x - 5y^4 - 7x$

Add or subtract.

9. $(2x^2 + y) + (x^2 - y)$

10. $(4xy - 7) + (3xy + y)$

11. $(y^4 + y + 1) - (y^4 - 2y + 6)$

12. $(x^3 + x) - (1 + x)$

Multiply.

13. $(2)(xy)$

14. $(x)(x)$

15. $(2y^2)(3y)$

16. $(-6xy^3)(2xy^2)$

17. $2(x + y + 3)$

18. $x(x^2 + 1)$

19. $-3y(y^2 + 2y + 1)$

20. $2xy(x^2 - y^2)$

21. $(x + 1)(x + y + 1)$

22. $(y - 1)(y^2 - 2y + 3)$

Multiply by FOIL.

23. $(x + 1)(x + 8)$ 25. $(2x + 6)(x + 9)$
24. $(y - 3)(y - 7)$ 26. $(3x - 4y)(2x - y)$

Multiply by the Difference of Squares.

27. $(x + 3)(x - 3)$ 29. $(2x + 3y)(2x - 3y)$
28. $(y - 10)(y + 10)$ 30. $(A + B)(A - B)$

Square the binomials.

31. $(x + 4)^2$ 33. $(2x - 3y)^2$

32. $(y - 5)^2$ 34. $(A + B)^2$

Chapter 7: List the factors of each.

1. $x \cdot x = x^2$ 3. $(x + 1)(x + 2) = x^2 + 2x + x + 2$
2. $x(x + 5) = x^2 + 5x$ 4. $y^2 - 5y + 6 = (y - 2)(y - 3)$

Find the GCF.

5. 12 , 18 8. $x(x + 3)$, $(x + 3)$
6. x^4 , x^3 , x^2 9. xyz , xy , yz
7. xy^2 , x^2y 10. $2p^2$, $3p$, 2

Fill in the missing factors.

11. $10 = (2)(\ \ )$ 15. $9(\ \ +\ \ ) = 9 \cdot 4 + 9 \cdot 5$
12. $8y = (\ \ )(2)$ 16. $9(\ \ \ \ ) = 9x + 18$
13. $x^6 = (x^4)(\ \ )$ 17. $3(\ \ \ \ ) = 3x - 24$
14. $y = (\ \ )(y)$ 18. $-4(\ \ \ \ ) = -4z + 40$

Factor out the GCF.

19. $3x^3 + 6x + 9$ 21. $7x + 7y + 7z$
20. $x^3 + 5x^2 + x$ 22. $(x + 2) \cdot x + (x + 2) \cdot 3$

Factor by grouping.

23. $x^2 + 2x + 3x + 6$

Factor these trinomials.

24. $x^2 + 7x + 12$
25. $3y^2 - 5y + 2$
26. $2x^2 + 14x + 20$
27. $x^2 + 3xy + 2y^2$
28. $y^2 + 10y + 2$

Factor these binomials.

29. $x^2 - 25$
30. $y^2 - 100$
31. $7y^2 - 28$
32. $y^2 + 1$
33. $4y^2 - 25$

Solve these equations

34. $x^2 + 7x + 12 = 0$
35. $7y^2 - 28 = 0$
36. $x^2 - 3x = 0$

37. $2x^2 + 15x = -7$
38. $x^2 = 3x - 2$
39. $14x^2 = x$

Cumulative Review

Simplify.

    1. $x + 5y - 5y$ 3. $-(x - 3x + 2)$ 5. $-3(x + 2y) + x + y$

    2. $-2 + 7x + 10$ 4. $7 + 2(y + 4)$ 6. $10 - (4p + 5) + 2p$

Solve.

    7. $x + 5 = 9$ 11. $9(p + 1) = 8p - 1$

    8. $5x = 10$ 12. $-(x - 1) = 1$

    9. $4x - 7 = 13$ 13. $2(x - 6) + x = 12$

    10. $4a = 2a + 10$ 14. $\frac{1}{6}x - 2 = \frac{2}{3}$

Solve.

    15. $-2p \le -8$ 16. $6x - 9 \le 0$

Simplify.

    17. $2^3 \cdot 2^{10}$ 18. $(2^3)^{10}$ 19. $\frac{2^{10}}{2^3}$ 20. x^{-3} 21. $(\frac{7}{x})^3$

Simplify.

    22. $7x + 3y - x + 2 + 8$

    23. $(4y - 3x) - (2 + y - x)$

    24. $x^2 + 3x + 1 + 3x^2 + x + 1$

Multiply.

    25. $(4x)(2x^2)$ 27. $x(x^2 + 2x + 1)$ 29. $(7x - 2)(x + 1)$

    26. $-3(x - y + 1)$ 28. $(x + 1)(x + 1)$ 30. $(8y + 1)(2y + 4)$

Factor.

    31. $x^2 - 4$ 34. $2y^2 + 7y + 3$

    32. $x^2 - 4x + 4$ 35. $4x^2 - 4$

    33. $3x + 3y + 3z$ 36. $6y^2 - 8y + 2$

Solve.

37. $x^2 + 2x = 0$ 38. $-x^2 + 9 = 0$ 39. $2y^2 + 7y + 3 = 0$

Answers

Review: Pre-Algebra, Page 5.

1. Commutative addition
2. Commutative multiplication
3. Associative addition
4. Associative multiplication
5. Identity addition
6. Identity multiplication
7. Commutative addition
8. Associative addition
9. Identity multiplication
10. Associative multiplication
11. Identity addition
12. Commutative multiplication
13. Commutative addition
14. Associative addition
15. Identity addition
16. Commutative multiplication
17. Associative multiplication
18. Identity multiplication
19. -4, 2, -11, 30, 0

20. -12
21. -5
22. 1
23. -4
24. -7
25. 5
26. 4
27. 3
28. -11
29. 11
30. -3
31. 18
32. -10
33. -11
34. -2
35. 16
36. 16
37. 31
38. 47
39. 37
40. 0

Distributive Law, Exercise 1, Page 6.

1.
Left Side	Right Side
7(2 + 3)	7 · 2 + 7 · 3
7(5)	14 + 21
35	35

2. 8(3) = 24   8 + 16 = 24
3. Both sides equal 900
4. 10
5. 9
6. 20
7. 20
8. 6

Distributive Law, Exercise 2, page 6.

1. 4 · 3 + 4 · 7
2. 6 · 1 + 6 · 2
3. 5 · 5 + 5 · 4
4. 9 · x + 9 ·2
5. 7 · y + 7 · 1
6. 2 · 3 + 2 · m
7. 5 · x + 5 · y
8. 3(-1) + 3 · 6

9. 12 · (-2) + 12 · 3
10. 4(-x) + 4 · 3
11. 7 · 3 + 2 · 3
12. 1 · 2 + 6 · 2
13. -2 · 4 + 7 · 4
14. -12 · 7 + 5 · 7
15. x · 4 + 3 · 4
16. a · b + a · c

Distributive Law, Exercise 3, page 8.

1. Terms = -5x, -3x, -2y, -7y
   Coefficients = -5, -3, -2, -7
   Like terms = -5x, -3x. Also -2y, -7y.
2. Terms = 7y, -2x, 14x
   Coefficients = 7, -2, 14
   Like terms = -2x, 14x
3. Terms = 4z, x, -6x
   Coefficients = 4, 1, -6
   Like terms = x, -6x
4. Terms = 2x, 3y, 4z, 5
   Coefficients = 2, 3, 4, 5
   Like terms = none

5. Terms = 10x, -y
   Coefficients = 10, -1
   Like terms = none
6. Terms = x and - x
   Coefficients = 1 and -1
   Like terms = x and - x
7. Terms = 9 and -6
   Coefficients = 9 and -6
   Like terms = 9 and -6
8. Terms = -89x
   Coefficients = -89
   Like terms = none
9. Terms = 4x, 5, 8
   Coefficients = 4, 5, 8
   Like terms = 5, 8.
10. Terms = -3, 2, 0, z
    Coefficient = -3, 2, 0, 1
    Like Terms = -3, 2, 0

Distributive Law, Exercise 4, page 9.

1. 9x
2. 10y
3. 11d
4. 0y or 0
5. -2x + 7y
6. $\frac{2}{3}$x
7. 2x + 5
8. 10y

9. 0
10. 10x + 3y + 2p (any order)
11. -13x - 6y
12. 4.00x
13. 7 - 4x or -4x + 7
14. -a + 9 or 9 - a
15. 11a + 14x + 7y (any order)
16. 13x

Distributive Law, Exercise 5, page 10.

1. -2x + 6
2. -2y - 18
3. 3z - 3
4. -14y + 21
5. -4x - 8
6. 15x + 6 Because 5x · 3 = 15x,
   and 2 · 3 = 6.
7. 10m - 25
8. 8x - 12

9. 18z + 12
10. x + 10
11. -14x + 7
12. 4x + 12y + 4
13. 6y - 10p - 8
14. -5a - 10b + 30
15. 3x - 6y + 21
16. -15x + 10

Distributive Law, Exercise 6, page 11.

1. -1 · 7
2. -1· 93
3. -1y
4. -1(x + 9)
5. -1(x - 6)

6. -1(y + 2x - 3)
7. 5 - 1(x + 4)
8. 7x - 1(3y + x)
9. 4 - 1(x + 2) - 1(y - 2)

Distributive Law, Exercise 7, page 12.

1. Method 1: -(-2x + 5) = -1(-2x + 5) = 2x - 5
   Method 2: -(-2x + 5) = 2x - 5
2. -5y + 6
3. -7y + 4
4. -x - 2y + 6z + 9
5.
Method 1	Method 2
7 - (x + 2y)	7 - (x + 2y)
7 - 1(x + 2y)	change signs
7 - x - 2y	7 - x - 2y

6. 5x + 2y + 1
7. 14 - 2x + 6p - 4a + q
8. -x + y - 3 + 4z. 4z remains positive,
   because it was not in parentheses.

Distributive Law Exercise 8, page 13.

1. $2x + 2$ 9. -2
2. $-3y - 12$ 10. $2 - 2y$ or $-2y + 2$
3. $7x + 2$ 11. $-6y$
4. -8, since $-2x + 2x = 0$.
5. $10y$ 12. $-z - 3$
6. $5p + 7$ 13. $\frac{1}{2}x + 13$, any order.
7. $-4x - 7y$ 14. $5x + \frac{1}{3}$ Hint: $\frac{1}{3} \cdot \frac{6}{1} x = 2x$.
8. $3x - 2$ 15. $-\frac{1}{2} + x$
                16. $7a + 2b + 11$

Linear Equations, Exercise 1, page 14.

1. 41 Hint: $-(-4) = +4$ 4. a. 11
2. 20   b. -4
3. a. $2(-5) + 11 = 1$ 5. a. $-(-2) + 7 = 9$
   b. 31   b. 8
   c. 11   c. 0

Linear Equations, Exercise 2, page 15.

1. Yes   5. Yes
2. No   6. Yes
3. Yes   7. No
4. No   8. Yes Hint: $4 = \frac{4}{1}$

Linear Equations, Exercise 3, page 17.

1. $x = 7$ Check $(7) - 3 = 4$ 7. 125
2. $y = -1$ 8. 5
3. 11   9. 2
4. 3   10. $z = \frac{4}{2} = 2$
5. 7   11. - 20
6. -23   12. 0

Linear Equations, Exercise 4, page 20.

1. a. $y = 2$ f. -4
     Check $5(2) = 10$ g. $\frac{10}{3}$
   b. $x = -4$ h. 1
   Hint: Divide by -3   i. 0
   c. $\frac{5}{2} = y$ j. -1
   Hint: Divide by 2   k. 1
   d. $-\frac{3}{4}$ l. $-\frac{1}{2}$
   e. $\frac{2}{10} = \frac{1}{5}$
2. See above.
3. a. $x = -1$ c. -4
   b. $y = 7$ d. 0
4. a. $x = 12$ c. $6 = y$
   b. $y = 2$ d. - 24
     $(\frac{2}{3} \cdot \frac{3}{1} = 2)$

Linear Equations, Exercise 5, page 22.

1. $x = 2$ 8. 3
   Check: $5(2) - 7 = 3$ 9. 2
          $10 - 7 = 3$ 10. 2

2. $y = 10$ 11. - 1
3. $x = -3$ 12. - 3
4. 7   13. - 2
5. - 11   14. 2
6. $\frac{3}{4}$ 15. 0
7. 7   16. - 2

Linear Equations, Exercise 6, page 24.

1. a. LCM = 9   4. $x = 2$, Hint: $\frac{4}{1} \cdot \frac{1}{4}x + \frac{4}{1} \cdot \frac{1}{2} = 4$
   b. $x = -12$, $x + 2 = 4$
  Hint: $\frac{9}{1} \cdot \frac{1}{9}x + \frac{9}{1} \cdot \frac{2}{1} = \frac{9}{1} \cdot \frac{2}{3}$
            $x + 18 = 0$ 5. a. After distributing 3,
                              LCM = 2.
2. a. LCM = 10   b. $x = 1$
   b. $x = 40$ 6. a. LCM = 4
3. a. LCM = 12   b. $x = -\frac{29}{4}$
   b. $x = -\frac{2}{3}$

Number Lines, Exercise 1, page 26.

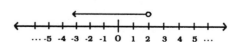

Number Lines, Exercise 2, page 26.

1. $5 < 10$ 6. >   11. >
2. >   7. <   12. <
3. <   8. <   13. <
4. <   9. >   14. >
5. <   10. <   15. >

Number Lines, Exercise 3, page 27.

1. True   4. True   7. True
2. True   5. True   8. True
3. False   6. False   9. False

Number Lines, Exercise 4, page 27.

1. a. $6 - 5 \geq -7$ b. $-3 - 5 \geq -7$
      $1 \geq -7$ True $-8 \geq -7$ False
2. a. $x \geq 3$

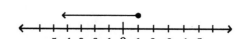

   b. $x < 2$

   c. $x \leq 1$

d. $x > -2$

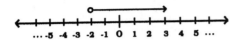

e. $p \le 0$

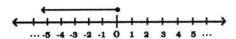

f. $x \ge -4$

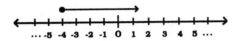

g. $y < 1$

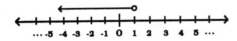

h. $z > 0$

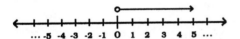

Number Lines, Exercise 5, page 29.

1. $x \ge -5$ 7. $x \le -\frac{3}{7}$

2. $x < 7$ 8. $x \le 3$

3. $y < 2$ 9. $y \le 1$

4. $x > -2$ 10. $x < 3$

5. $p \ge \frac{1}{2}$ 11. $x \ge \frac{5}{2}$ or $\frac{5}{2} \le x$

6. $x \le -6$ 12. $a < \frac{7}{4}$

Review, Chapter 2 - 4, page 30.

Chapter 2.

1. Terms: $-3x, 2y, -x, 19, -4$
   Coefficients: $-3, 2, -1, 19, -4$
   Like terms: $-3x, -x$, also: $19, -4$
2. $-4x + 2y + 15$
3. a. $-12y + 8x - 4$ c. $-x + 2$
   b. $42p - 7$ d. $-y + 4x - 1 + p$
4. a. $2x + 7$ d. $2a + 6$
   b. $-3p + 8$ e. $y - 3$
   c. -12 f. $2x + 4y$

Chapter 3

1. a. Yes   c. No
   b. Yes   d. Yes

2. a. 11   e. 5   i. -6
   b. 4   f. -2 j. 0
   c. 5   g. $\frac{3}{4}$ k. 18
   d. 0   h. $\frac{7}{5}$ l. $-\frac{1}{6}$
3. a. -3 e. 9   i. 2
   b. -5 f. -3 j. 1
   c. $\frac{5}{2}$ g. 0   k. -1
   d. -2 h. 1   l. 5
4. a. $\frac{8}{3}$ c. 16
   b. $-\frac{3}{2}$

Chapter 4

1. a. $<$ b. $<$ c. $<$
2. a. Yes   b. Yes   c. No
3. a. $y > -1$

b. $p \le -2$

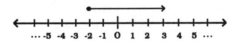

c. $x \ge -2$

d. $x < 3$

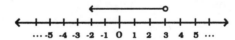

Prop. of Exponents, Exercise 1, page 33.

1. 2   4. $\frac{1}{x}$ 7. 1

2. $\frac{1}{5}$ 5. $a \cdot a \cdot a$ 8. b

3. y   6. $\frac{1}{y}$ 9. $\frac{1}{y \cdot y}$

Prop. Exponents, Exercise 2, page 34.

1. $4 \cdot 4 \cdot 4$ 11. 3^3
2. $\frac{1}{3} \cdot \frac{1}{3}$ 12. 4^2
3. $(-4)(-4)(-4)(-4)(-4)$
4. $x\,x\,x\,x\,x$ 13. x^6
5. $y\,y\,y\,y\,y\,y$ 14. x^3
6. y   15. 7^1
7. $z\,z\,z$ 6. y^1
8. $(2y)(2y)(2y)$ 17. p^5
9. $(3b)(3b)(3b)(3b)$ 18. $(2a)^3$
10. $(x^2)(x^2)(x^2)$ 19. $(x^2)^3$

Prop. of Exponents, Exercise 3, page 36.

1. x^7
2. x^{10}
3. x^3
4. x^{10}
5. x^{21}
6. x^{21}
7. x^4
8. y^{40}
9. y^{14}
10. 2^7
11. 2^{12}
12. 2^1 or 2
13. x^8 Hint: $x = x^1$
14. 6^9
15. x^7
16. 4^3
17. y^2
18. y
19. 5^{16}
20. 5^2
21. 5^8
22. y^2
23. x^2
24. x^0 or 1

Prop. of Exponents, Exercise 4, page 37.

1. x^2y^2
2. $2^3y^3 = 8y^3$
3. $16y^6$
4. $4^{15}y^5$
5. $3^8a^2b^2$
6. $a^4x^{16}y^8$
7. $\frac{4}{9}$
8. $\frac{x^3}{27}$
9. $\frac{64}{y^9}$
10. $\frac{49}{x^2y^{10}}$
11. $\frac{49}{x^2y^6}$
12. $\frac{125a^9}{b^9}$

Prop. of Exponents, Exercise 5, page 38.

1. $\frac{1}{x^7}$
2. $\frac{1}{x^5}$
3. $\frac{1}{y^2}$
4. $\frac{1}{y^{12}}$
5. $\frac{1}{x^{73}}$
6. $\frac{1}{2^4}$ or $\frac{1}{16}$
7. $\frac{1}{3^3}$ or $\frac{1}{27}$
8. $\frac{1}{25}$
9. $\frac{1}{x^1} = \frac{1}{x}$
10. $\frac{1}{y}$
11. $\frac{1}{5}$
12. $\frac{1}{3x}$

Prop. of Exponents, Exercise 6, page 39.

1. 1
2. 1
3. 1
4. 1
5. 1

Prop. of Exponents, Exercise 7, page 41.

1. x^{12}
2. x^4
3. x^{-4} or $\frac{1}{x^4}$
4. $\frac{1}{x^{10}}$
5. x^2
6. x^{10} see last two examples
7. $x^{-10} = \frac{1}{x^{10}}$
8. $\frac{1}{x^2}$
9. $\frac{1}{y^6}$
10. y^{12}
11. $\frac{1}{3^3}$
12. $\frac{1}{x^2}$
13. x^{-2} or $\frac{1}{x^2}$
14. $\frac{1}{x}$
15. y^4
16. $x^2 \frac{1}{y^4}$
17. 3^8y^{12}
18. $\frac{x^2}{y^6}$
19. 1
20. x^4

Hint: $\frac{x}{x^3} = \frac{x^1}{x^3} = x^{1-3}$

Polynomials, Exercise 1, page 42.

1. Degree of : $3y^2$ is 2; $-y$ is 1; 4 is 0.
   Polynomial degree is 2.
2. Degree of: $2x^5$ is 5; $-x$ is 1.
   Polynomial is 5.
3. x^2y^5 is 7; $2x^7y$ is 8.
   Polynomial is 8.
4. x^2 is 2; y^4 is 4; 9 is 0.
   Polynomial is 4.
5. $4a^2b^3$ is 5; $8a^1b^4$ is 5.
   Polynomial is 5.
6. $x^2y^3z^5$ is 10; 8 is 0.
   Polynomial is 10.
7. $x^{10} + x^2$
8. $x^9 + x^3$
9. $4y^4 + y^2 + 3$
10. $8y^8 + 6y + 5$
11. $x^3 + x^2 + x + 3$

Polynomials, Exercise 2, page 43.

1. Yes, polynomial; binomial.
2. No, not polynomial.
3. Yes, monomial.
4. No.
5. Yes, trinomial.
6. Yes, monomial.
7. Yes.
8. Yes, trinomial.

Polynomials, Exercise 3, page 44.

1. $7x^2$
2. $3x^5$
3. $6y^7 - 2x$
4. $7x^3 + 4x$
5. $5y^2 + 6y + 7$
6. $-2x^2 - 19x$
7. $6x^6y + 5$
8. $4xy^5 + 4x$
9. $y^7 - 3y^4 + 2y$
10. $a^3 b$
11. $9xy^5 + 2x^5y$
12. $6 - 3b$

Polynomials, Exercise 4, page 45.

1. $7x + 2$
2. $15x^7 - 8y$
3. $7y^2 + 3y$
4. x^2
5. $2y^3$
6. $-4xy + 4x$
7. $3x + 11$
8. $3x^{15} - 5y$
9. 0
10. $2xy^2 + 3$
11. $-y^2 - 3y$
12. $11x^2y - 7x - 3$

Polynomials, Exercise 5, page 46.

1. $15x^6$
2. $-70y^{12}$
3. $12x^4$
4. $6x^7y^3$
5. $-5x^8y^3$
6. $2x^2$
7. $3y^2$
8. a^2
9. $9x^{40}y^{40}$
10. $-14x^2y$
11. $-xy^5$
12. xy^2z Any order.
13. $2p^3q^3$
14. $-8x^2$

Polynomials, Exercise 6, page 47.

1. $7x + 14$
2. $3y^2 - 15y$
3. $x^2 + 2x$
4. $2y^2 - 8y$
5. $4x^3 + 4x^2$
6. $6y^5 - 18y^3$
7. $5x^6 + 10x^2$
8. $y^3 - y^2 + 2y$
9. $a^5 + a^3 + a^2$
10. $x^2y - xy^2$
11. $x^5 + 2x^2y + xy$
12. $3x^3y + 2x^2y^2$
13. $4x^3y^2 + 8xy^2$
14. $7x^6y + 7x^5y$
15. $-3p^3q + p^2q^2$

Polynomials, Exercise 7, page 49.

1. $10x^3 \underline{+ 2x^2} + 8x \underline{- 15x^2} - 3x - 12 =$
   $10x^3 \underline{- 13x^2} + 5x - 12$.
2. $6y^3 + \underline{3y^2} - 12y + \underline{2y^2} + y - 4 =$
   $6y^3 + \underline{5y^2} - 11y - 4$
3. $7x^3 - 4x^2 - 5x + 2$
4. $a^3 - 4a^2 - 4a + 1$
5. $10x^2 + 2x + 15x + 3 = 10x^2 + 17x + 3$

Polynomials, Exercise 8, page 51,

1. $8x^2 - 2x - 15$
2. $6y^2 + 7y - 20$
3. $x^2 + 5x + 6$
4. $y^2 + 3y - 10$
5. $5y^2 - 18y - 8$
6. $x^2 + 6x + 5$
7. $p^2 + 4p - 12$
8. $2x^2 - 7x + 6$
9. $y^2 - 7y + 12$
10. $x^2 + 2x + 1$
11. $7x^2 + 14x + 7$
12. $3y^2 - 7y + 2$
13. $p^2 - 4$ Hint: $-2p + 2p = 0$
14. $4x^2 + 8x + 3$
15. $10y^2 - 9y + 2$
16. $2y^2 + 6y + 4$
17. $5p^2 - 20p + 20$
18. $7y^2 - 7y - 42$
19. $x^2 + 2xy + y^2$
20. $x^2 - y^2$ Hint: $-xy + xy = 0$
21. $x^2 + 3xy + 2y^2$
22. $3a^2 + 7ab + 2b^2$

Polynomials, Exercise 9, page 52.

1. $x^2 - 81$
2. $y^2 - 100$
3. $4x^2 - 1$
4. $a^2 - b^2$
5. $x^2 - 1$
6. $p^2 - 49$
7. $p^2 - 49$
8. $9x^2 - 16$
9. $16y^2 - 25$
10. $x^2 - 16y^2$
11. $x^4 - 121$
12. $A^2 - B^2$

Polynomials, Exercise 10, page 53.

1. $x^2 - 6x + 9$
2. $y^2 + 18y + 81$
3. $4x^2 + 4x + 1$
4. $A^2 + 2AB + B^2$
5. $x^2 - 14x + 49$
6. $x^2 + 14x + 49$
7. $16y^2 + 40y + 25$
8. $x^2 + 10xy + 25y^2$
9. $x^6 + 22x^3 + 121$
10. $A^2 - 2AB + B^2$

Factoring, Exercise 1, page 56.

1. Product 77, factors 7 and 11.
2. Product x^8, factors x^5 and x^3.
3. Product $4y - 12$, factors 4 and $(y - 3)$.
4. Product $x^2 - 2x + 6x - 12$, factors $x + 6$ and $x - 2$.
5. Product 42, factors 7 and 6.
6. Product $3x + 2$, factors 3 and $(x + 2)$.
7. Product $y^2 - 25$, factors $(y + 5) (y - 5)$.
8. Product $x^2 + 3x + 1$, factors $(2x + 1) (x + 1)$.
9. Learn how to find factors when only product is known.

Factoring, Exercise 2, page 57.

1. 6
2. 14
3. 1
4. x^2
7. xy
8. pq^2
9. 1
10. 4x
13. 3
14. 1
15. 4
16. $2(x - 2)$

5. y^4
6. 1
11. $3y^2$
12. $5x^2y$
17. $2y - 1$
18. $x(x + 3)$

Factoring, Exercise 3 page 59.

1. 5
2. -5
3. 2x
4. x
5. y
6. x
7. y^4
8. 2x
9. $2y^2$
10. 5a
11. $2pq^2$
12. $-x^4$
13. $5(8 + 9)$
14. $3(11 + 2)$
15. $4(x + 8)$
16. x, 3, $7(x+3) = 7x + 21$
17. y, -2, $-3(y - 2)$
18. x, 10, $x(x + 10)$
19. $8(x + 2)$
20. $-4(y + 5)$
21. $x(y + 2)$

Factoring, Exercise 4, page 61.

1. $4(x + 3)$
2. $2(x - 5)$
3. No common factor.
4. $6(x - 1)$
5. $3x(x + 2)$
6. $5y^2(1 - 2y)$
7. No common factor.
8. $p^3(7p^3 - 3)$
9. $2x(2x^3 + 1)$
10. $2(x^2y^2 + 2x^2 - 3x + 5)$
11. $5(x + y + z)$
12. $3(x^2 - xy + y)$
13. $a(x + y + z)$
14. No common factor.
15. $(x - 1) (x + y + 3)$
16. $(y + 1) (xy + 4x - 5)$
17. $(x - 3) (x + 2)$
18. No common factor.
19. $(x + 2) (x + 2)$
20. $(x - 3) (x + 1)$
Hint: $x(x - 3) + x - 3 =$
$x(x - 3) + 1(x - 3)$

Factoring, Exercise 5, page 63.

1. $(y + 2) (x + 3)$
2. $(x - 5) (x + 4)$
3. $(y + 7) (x - 2)$ See Example 2.
4. $(3y - 4) (y - 3)$
5. $5(x + 3)(y + 2)$
6. $4(y + 5) (y + 5)$
7. Does not factor.

Factoring, Exercise 6, page 67.

1. $(x + 2) (x + 3)$
2. $(y - 3) (y - 4)$
3. $(2x + 1) (x + 2)$
4. $(5y - 1) (y + 2)$
5. Not factorable
6. $(2x - 1) (2x - 1)$
7. $(2x + 3) (2x + 5)$
8. $2(x - 3) (x - 4)$
9. $3(3x + 1) (2x + 1)$
10. $3p(p + 2) (p + 1)$
11. $(x + 7y) (x + 7y)$
12. $(x - 5y) (x + 7y)$
13. $(2x + y) (x + y)$
14. $(3p + q) (p + q)$
15. Not factorable
16. $(7a - b) (a - 2b)$
17. $5(x + y) (x + y)$
18. $7(x - 2y) (x - y)$
19. $2(4x + y) (x + y)$
20. $x(x + 3y) (x + 3y)$

Factoring, Exercise 7, page 69.

1. $(x + 4) (x - 4)$
2. $(y + 8) (y - 8)$
3. $(x + 7) (x - 7)$
4. $3(y + 3) (y - 3)$
5. $x(y + 2) (y - 2)$
6. $7y(x + 10) (x - 10)$
7. $x(x + y) (x - y)$
8. Not factorable.
9. $(x + 1) (x - 1)$
10. $7(x + 1) (x - 1)$
11. $(8x + 5) (8x - 5)$
12. $(3x + 5) (3x - 5)$
13. $5(3x^3 + 2y^2) (3x^3 - 2y^2)$
14. $(5x^2y + 3pq) (5x^2y - 3pq)$
15. Not factorable

Factoring, Exercise 8, page 69.

1. $(x + 1)(x + 1)$
2. $(x + 3)(x - 3)$
3. $(x + 2)(x + 3)$
4. $3(y + 4)(y - 4)$
5. $(2x + 1)(x + 1)$
6. $3(y + 1)(y + 1)$
7. $5(x + 2)(x - 2)$
8. $2(2y^2 + 3y + 6)$
9. $(3y + 2)(y + 1)$
10. Not factorable
11. $y(y + 3)(y - 3)$
12. Not factorable
13. $y(x^2 - 4x + 2)$
14. $x(x - 3)(x - 1)$
15. $2(y^2 - 3)$
16. $5(x + 7)(x + 2)$
17. $7(x^2 - 7x + 2)$
18. $a(b + 1)(b - 1)$

Factoring, Exercise 9, page 72.

1. $x = -\frac{1}{2}$ or $x = -1$
2. 3, 4
3. $0, -\frac{1}{4}$
4. 3, -3
5. $-1, \frac{-3}{2}$
6. 0, -3
7. 2, 3
8. -3
9. - 5, 2
10. 1, 4
11. 0, -2
12. 5, -5

Review Chapters 5 - 7, page 73.

Chapter 5.

1. x^{11}
2. z^{18}
3. y^9
4. 1
5. $\frac{1}{x^5}$
6. x^4y^4
7. y^5
8. z
9. 2^{20}
10. 4
11. 1
12. $\frac{1}{x^9}$
13. $\frac{1}{y^5}$
14. $\frac{1}{x^{10}}$
15. $\frac{1}{y^4}$
16. z^6

Chapter 6

1. Binomial, degree =5
2. Trinomial, degree = 1
3. Monomial, degree = 2
4. Binomial, degree = 6
5. $x^2 + 8x - 1$
6. $-5x - 6$
7. $-3yx^2 + z$
8. 0
9. $3x^2$
10. $7xy - 7 + y$
11. $3y - 5$
12. $x^3 - 1$
13. $2xy$
14. x^2
15. $6y^3$
16. $-12x^2y^5$
17. $2x + 2y + 6$
18. $x^3 + x$
19. $-3y^3 - 6y^2 - 3y$
20. $2x^3y - 2xy^3$
21. $x^2 + xy + 2x + y + 1$
22. $y^3 - 3y^2 + 5y - 3$
23. $x^2 + 9x + 8$
24. $y^2 - 10y + 21$
25. $2x^2 + 24x + 54$
26. $6x^2 - 11xy + 4y^2$
27. $x^2 - 9$
28. $y^2 - 100$
29. $4x^2 - 9y^2$
30. $A^2 - B^2$
31. $x^2 + 8x + 16$
32. $y^2 - 10y + 25$
33. $4x^2 - 12xy + 9y^2$
34. $A^2 + 2AB + B^2$

Chapter 7

1. x,x
2. x, x + 5
3. x + 1, x +2
4. y - 2, y - 3
5. 6
6. x^2
7. xy
8. x + 3
9. y
10. 1
11. 5
12. 4y
13. x^2
14. 1
15. $9(4 + 5)$
16. $9(x + 2)$
17. $3(x - 8)$
18. $-4(z - 10)$

19. $3(x^3 + 2x + 3)$
20. $x(x^2 + 5x + 1)$
21. $7(x + y + z)$
22. $(x + 2)(x + 3)$
23. $(x + 2)(x + 3)$
24. $(x + 3)(x + 4)$
25. $(3y - 2)(y - 1)$
26. $2(x + 2)(x + 5)$
27. $(x + 2y)(x + y)$
28. Not factoravle
29. $(x + 5)(x - 5)$
30. $(y + 10)(y - 10)$
31. $7(x + 2)(x - 2)$
32. Not factorable
33. $(2y + 5)(2y - 5)$
34. -3, -4
35. 2, -2
36. 0, 3
37. $-\frac{1}{2}, -7$
38. 2, 1
39. $0, \frac{1}{14}$

Cumulative Review, page 76.

1. x
2. $7x + 8$
3. $-x + 3x - 2$
4. $2y + 15$
5. $-2x - 5y$
6. $-2p + 5$

7. $x = 4$
8. 2
9. 5
10. 5
11. -10
12. 0
13. 8
14. 16

15. $p \geq 4$
16. $x \leq \frac{3}{2}$

17. 2^{13}
18. 2^{30}
19. 2^7
20. $\frac{1}{x^3}$
21. $\frac{7^3}{x^3}$

22. $6x + 3y + 10$.
23. $3y - 2x - 2$
24. $4x^2 + 4x + 2$

25. $8x^3$
26. $-3x + 3y - 3$
27. $x^3 + 2x^2 + x$
28. $x^2 + 2x + 1$
29. $7x^2 + 5x - 2$
30. $16y^2 + 34y + 4$

31. $(x + 2)(x - 2)$
32. $(x - 2)(x - 2)$
33. $3(x + y + z)$
34. $(2y + 1)(y + 3)$
35. $4(x + 1)(x - 1)$
36. $2(3y - 1)(y - 1)$

37. 0, -2
38. 3, -3
39. $-\frac{1}{2}, -3$

English Series

The Straight Forward English Series is designed to measure, teach, review, and master specific English skills: capitalization and punctuation; nouns and pronouns; verbs; adjectives and adverbs; prepositions, conjunctions and interjections; sentences; clauses and phrases; and mechanics.

Each workbook is a simple, straightforward approach to learning English skills. Skills are keyed to major school textbook adoptions.

Pages are reproducible.

GP-032 Capitalization and Punctuation
GP-033 Nouns and Pronouns
GP-034 Verbs
GP-035 Adjectives and Adverbs
GP-041 Sentences
GP-O43 Prepositions, Conjunctions, & Interjections

ADVANCED SERIES, large editions
GP-055 Clauses & Phrases
GP-056 Mechanics

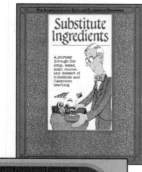

Substitute Teaching

GP-027 Substitute Teacher Folder
A pertinent information folder left by regular classroom teachers listing class schedules, classroom procedures, discipline, support personnel, and regular classroom teacher expectations.

GP-001 Substitute Ingredients
A collection of imaginative language arts, math, and art activities for grades 3–8. Reproducible master sheets accompany most lessons.

GP-002 Mastering the Art of Substitute Teaching
Substitute teaching formats, strategies, and activities strictly from practical experience.

GP-003 Classroom Management for Substitute Teachers
Suggested procedures for being-in-charge, establishing rapport, and getting the support of regular classroom teachers and staff.

GP-014 Lesson Plans for Substitute Teachers
A packet of 12 lesson plan forms to be filled out by regular classroom teachers to provide one day of instruction during their absence.

GP-004 Just Fun
Engaging, high-interest activities that are short span, 10-15 minutes in length.